CLIMATE POLITICS ON THE BORDER

RHETORIC, CULTURE, AND SOCIAL CRITIQUE

Bob's Collab

CLIMATE POLITICS ON THE BORDER

ENVIRONMENTAL JUSTICE RHETORICS

KENNETH WALKER

THE UNIVERSITY OF ALABAMA PRESS   TUSCALOOSA

The University of Alabama Press
Tuscaloosa, Alabama 35487-0380
uapress.ua.edu

Typeface: Scala Pro

Cover image: Details from *De Todos Caminos Somos Todos Uno*,
2018, by Adriana M. Garcia, mural on display at San Pedro Creek
Culture Park, commissioned by Bexar County Commissioners
Court; photograph © Al Rendon photography

Cover design: Michele Myatt Quinn

Cataloging-in-Publication data is available from the Library of Congress.
ISBN: 978-0-8173-2111-6
E-ISBN: 978-0-8173-9384-7

Contents

Figures

Preface

Haunted Emergenc(i)es

> Three centuries of history is but a tick in the legendary Clock of the
> Long Now, yet time enough to have made of *San Antonio de Béjar*
> a veritable metropolis of ghosts. Indeed, perhaps there are more
> ghosts in San Antonio than there are living souls in our 21st century
> American city.
>
> JOHN PHILLIP SANTOS, *De Unos Lugares Perdidos*

AS A WORLD OF MANY WORLDS, San Antonio is an invitation toward living in
multiple dimensions. Its bordered thresholds are haunted by all of those
who have lived, loved, and lost in this city. These ghosts animate the city's
languages, landscapes, and ways of life; through the feedbacks, recircula-
tions, and resonances of more-than-human histories, futures, and presences,
these ghosts co-constitute this place. They amplify the city's rhythms. They
participate in its contemporary discourses, practices, and politics. Today migrant monarch butterflies will light in the front yards of this city, and migrant families will arrive at the bus station. Learning to notice these migrations is an embrace that induces me to think about generational migrations,
which, unimaginably, have also led me, and all my other near-inhabitants, to
these moments and these worlds through this place. This "legendary Clock
of the Long Now" contains stories deeper than the three centuries of this
self-fashioned colonial multiplex. These histories also co-produce futures of
travelers, migrants, and settlers who continually find themselves crafting a
life on these edges of the American Empire, the northern frontiers of Mexico, the mission town of Nueva España, the crossroads of many Indigenous
tribes, and in these fertile prairies now filled with the ghosts of bison, mast-
odons, saber-toothed cats, and flooded-out camarones (shrimp). This is indeed "nuestra querida ciudad de fantasmas, our beloved city of ghosts."[1]

If you are not afraid to make yourself vulnerable, if you allow these ghosts to embrace you, if you consider your inhabitation with them, they can induce you to think again from where you stand about the necessary politics of now. Like its ghosts, San Antonio's climate and weather exists outside, beyond, yet critically within this city's colonial social orders. Examining historical moments when climate and weather events exceed these structures, and make them radically apparent, is to think with a force that is always already otherwise, elsewhere, and yet in critical relation to the human and more-than-human. This book proposes that to engage with the politics of local climate and weather events is also to engage in a de/colonial praxis whose future will be increasingly haunted by an extractivist modern/colonial past that has created our contemporary climate breakdown.² This book narrates the emergenc(i)es of climate politics in *San Antonio de Béjar* in order to examine its haunted futures that may find its inhabitants slowly compelled to actions completely ill-equipped for the worldings to come.

in part, anticipatory

Una culebra de agua—this is how the Spanish described one of the earliest flooding events ever recorded in the heart of San Antonio in 1819, one hundred years after colonial contact. At the time, San Antonio was contested territory with barely two thousand inhabitants. The missionaries, soldiers, and local Coahuiltecan tribes who lived here were in constant conflict with the Comanches and Apaches who notoriously disrupted agrarian life and trade. When Napoleon sold Louisiana to the United States in 1803, San Antonio became a militarized site of revolutionary warfare among the Spanish, Mexican, and Republican Texian forces that destroyed San Antonio's emerging ranching economy. When the 1819 flood hit that July, it carried away stock, people, houses, and bridges and left the community in ruins.³ From colonial relations to land and people arise destructive colonial monsters, culebras de agua. To love these monsters is not to perpetuate relations of systemic domination but rather to attempt to create a different relation where land and people are co-constituted, where many worlds co-exist and at times come to rely on one another.

Tropes of flooding are multiply metaphorical. They function as analogical verifications of many social-material existences catalyzed by a flood event. In 1819, one analogue was political revolution—the revolutions in Texas and Mexico that precipitated Anglo-American immigration so Mexico could protect San Antonio from the Comanches and boost modern progress through the technologies of civilization.⁴ *Protection for progress.* San Antonio's culebra de agua was a different kind of revolution whose destructive force also fostered renewal and transformation. Likewise, Noah's flood intended to end corruption and debauchery, to renew and redeem its survivors. In *The Epic of Gilgamesh*, the flood is a catastrophe but also a preparation for a new world

to come. *Climate Politics on the Border* is about the force of relations between
extreme weather events and the inhabitants who live with them. At its core
is an interest in the character of the relations between a city's people and a
city's climate. What can San Antonio learn from the climate and weather
disturbances of its colonial past and present as these relations now oscillate
into its unprecedented climate futures? It asks San Antonians, and all city-
level climate politics, to engage its local histories in the future present tense
so one can imagine what we will have done tomorrow through the violent
colonial and climatic histories now on the horizon.

Although the 1819 flooding story comes from the first one hundred years
of Spanish colonial contact in San Antonio, it is worth remembering that at
the time it was the Spanish who had to seek entry into already existing Indigenous political economies and territorial disputes dominated by the Lipan Apache and Comanche.[5] Indeed, the subjugated bands of Coahuiltecan
tribes joined the Spanish missions in San Antonio less out of religious con-
version than by a pragmatic fear of the Apaches.[6] And even then, many individuals from these tribes traveled back and forth among the missions and
the various trails that created "dynamic Indigenous landscapes."[7] Such histories remind us that Indigenous cultures carry legacies of domination, exploitation, and inequality too. As Sandra Harding notes, "Indigenous and
non-Western cultures and their knowledge systems often legitimate cruel
and oppressive social relations, . . . [and] modern sciences and their philosophies provide attractive alternatives to such traditional knowledge systems."[8]
In the first century of colonial contact, Indigenous peoples actively negotiated Spanish colonialism, even if at various points, and especially toward
the end, all Indigenous tribes in South Texas suffered disease, displacement,
segregation, deculturation, and violent assimilation. The recorded flood stories from dominant cultures residing in San Antonio are haunted by the discourses and practices of those whose stories have been suppressed and lost
to human memory. What did "flooding" mean for the Payaya (and other Coahuiltecan tribes), Tonkawa, Lipan Apache, and Comanche? What figurations and analogical verifications did floods co-create with these people in
their thousands of years of living together?

Alvar Núñez Cabeza de Vaca was the first Spanish explorer to come in
contact with the inland Coahuiltecan tribes of South Texas in the sixteenth
century, but settler contact was not until June 13, 1691, when an entrada (ex-
ploratory expedition) commanded by Domingo Terán de los Ríos and Spanish Franciscan priest Damián Massanet first encountered the Indigenous
Payaya at a site near what is now San Pedro Springs—a lush and watery
plain with giant oak, cedar, willow, and cypress. Fray Massanet wrote: "In
the language of the Indians, it is called Yanaguana. . . . I called this place

San Antonio de Pádua, because it was his day."[9] Thus, the first act of colonial conquest was an act of bordered epistemologies—a bordered naming that yet privileged Western possession of a place in the cosmos ("it was his day"). Those acts of naming were followed by more religious, military, and colonial rituals seeking to impress and convert Indigenous tribes. The exact meaning of Yanaguana is unknown, yet we know the Payaya and their ancestors successfully inhabited this spring-fed landscape because of their ability to track seasonal migrations of animals and harvests of plants. We know the Payaya and similar bands carefully managed the landscapes and waterways of these territories, partially with the use of fire.[10] The place of Yanaguana is in the same location as San Antonio de Pádua, but it is also a place elsewhere and otherwise, yet still here in ghostly remnants.

What about these silences? What are the flood stories of the Payaya? Is this even the right question? Do nomadic peoples, who do not build immovable missions and presidios, conceptualize "flooding," which denotes "overflow" as flowing water submerges some determined boundary to engulf a fixed entity? At the time of contact, the river prairies were filled with large stretches of prairie grasses and luscious trees with creeks running through them. When it rains these plants take up that flowing water and there's less "runoff," even if at some point there is saturation. As a nomadic tribe forages for food, does the water "flood," or "rise" as you move to an encampment with higher elevation? In the Indigenous cosmologies of deep time, were these rising waters associated with destruction and creation? Were floods partial drivers in the impulse to migrate farther south where Mexica and Chichimeca codices note that the homeland, Aztlán/Chicomoztoc/Zuvuya, was a watery place, the hill of water, seven caves surrounding a lake.[11]

At the very least, in these dynamic Indigenous landscapes the relations between flowing water and communities of people allowed for flux and mobility in ways the Spanish colonizers completely rearranged and made stationary. Whatever the Payaya-language term is for flooding, what seems clear is just how much flash floods haunted the practices of Spanish colonial place-making, and have haunted San Antonio de Béjar at least since that time. These intersecting colonial legacies from the Spanish, Mexican, Texian, and American (United States), have been perpetually haunted by wrong relations to land, water, and climate that in turn precipitate wrong relations to co-inhabitants. And it is from these relations that colonialism describes flooding in monstrous terms as culebras de agua. From these wrong relations emerge colonial monsters.

I live, work, eat, pray, sleep, and raise my children in these river-threaded prairies, like the Indigenous, Spanish, northern European, and the rest of

Figure 1. Fray Antonio de Olivares's 1718 journal of baptisms on display in 2018 as part of the San Antonio tricentennial exhibition at the Witte Museum. Photograph by Kenneth Walker. Courtesy of the Catholic Archdiocese of San Antonio Archives.

the world that now arrives and crafts a life here in San Antonio. When Fray Antonio de Olivares founded the Mission San Antonio de Valero, he kept a sunburned leather-bound journal of baptisms. During the display of this artifact for San Antonio's three-hundred-year colonial anniversary, Andrés Tijerina, a historical consultant, explained, "It's called the book of baptisms, in his handwriting, and he names every person. And let me tell you something: Those are Indians, there's Spaniards, there's Mexicans. But you want the birth of the people of San Antonio? They were the Native Americans, and he's got who was born and what date!"[12] (see figure 1). Contrast this colonial contact with Jamestown, whose records do not name the Africans they brought with them in slavery, who were never given baptisms, and who were treated as chattel. The colonists of Jamestown brutally murdered Indigenous peoples, though for a time those tribes gave back as much as they got. To be sure, Indigenous peoples of what is now San Antonio didn't need to be baptized; they didn't need to dig acequias by hand. But not all colonial contact is equally violent, even if now we can unequivocally renounce it. All of us who

perhaps a better way of framing this is different types of violence.

live and breathe here must grapple with the ghosts of this legacy that is alive
and well and still haunting all of us who attempt to call this place home.

As anywhere, in San Antonio water tells many stories, and one of them is
that water was the primary resource for establishing colonial power relations
that by and large remain intact to this day. From Waco, Texas, to Monterrey,
Mexico, Central-South Texas and Northern Mexico is a place with magical
spring-fed water, a spring line. This spring line has offered watering holes,
ojos de agua, for hundreds of generations of travelers, Indigenous peoples,
immigrants, and settlers alike. As these immigrants became settlers they
built houses and businesses, and they learned to live with perpetual yet always surprising flash floods. Generations of capital, commerce, and colonial forces have had to contend with flash floods like bad dreams haunting
a restless sleep. From the eighteenth century to the present, during San Antonio's long era of coloniality, water was controlled and managed—for irri-
gating crops, for turning mills, for cleansing, for protection, for recreation,
and more. Along with these controls came logics of modernity and progress
that reified colonial relations of domination along the lines of race, class,
gender, sexuality, among others. Through the disturbances of flash flooding,
San Antonio mixed a western sense of water politics with a southern sense
of property rights as one of its unique assemblages of coloniality. Protection
for progress.

Floods. A rise of water. Overflows. Inundation. Submergence. But at
what defined level marks "over"? What excess of water is no longer bear-
able? What are the histories of these discharges? Limestone, karst, and escarpments certainly provide the momentum to create a "flash" flood, with
all its associations to lightning and thunderstorms, when rainfall and runoff
concentrate into a peak for a few minutes, or a few hours. There are many
narratives that San Antonians tell about flash floods, and they all mark a certain located point of view and practice. A common saying goes that flooding happens; it is people who provide the disaster. So, flooding happens and
people fashion lives filled with moments of beauty and ruin. Modern/colo-
nial development arose from the catastrophes, disturbances, and destruc-
tions of early flooding—disturbances that tended to reinforce inequitable
power relations. Yet, as Anna Tsing reminds us, "No single standard for assessing disturbance is possible; disturbance matters in relation to how we
live. . . . Disturbance is never a matter of 'yes' or 'no'; disturbance refers to
an open-ended range of unsettling phenomena."[13] But the modern/colonial
flood control systems in San Antonio seem to only say "no." No disturbance
here. This protection and progress has institutionalized precarity for generations of people that are still fighting for basic public services. Yet, floods also

fostered spaces and moments of profound creativity, poignant camaraderie, and radical coalitions.

It is difficult to fully anticipate what is to come for the next few generations of San Antonians. The auguries of climate science tell us it will get much hotter (maximums, averages, and duration of heat will all increase significantly over the next few decades), and we know San Antonio will suffer through more extreme droughts, and more flash flood events, with the likelihood of hundred-year storm events (over ten inches of rain in twenty-four hours) now twice as likely to occur.[14] A history of these events tells us that people will adapt to these changes within the structures of our modern/colonial world—and thus with everything that precipitates from them like disease, pollution, gentrification, and more. A changing climate means there will be more catastrophes, disturbances, and devastation, just as there will be more creativity, camaraderie, and coalitions. Floods will continue to stir the political cauldron.

The local history of climate politics I narrate in this book will face a future in which deliberations about shared sacrifice, public service equity, and urban developments will occur in a city facing at the very least more intensive heat waves and droughts, more intensive flooding events, and a massive influx of new people. By 2040, San Antonio will add another one million people to its population. At the same time, San Antonio will likely have at least twenty-four additional days each year of temperatures above 100 degrees.[15] What seems like a manageable climate shift will have profound consequences for San Antonio's quality of life—less time outdoors, higher utility bills, more heat-related illness and mortality, droughts, wildfires, and increased air pollution. Add to this the possibility that areas of the southern United States will see multiple catastrophic extreme weather events in just the next few decades.[16] It is not difficult to see how the displacement of people from New Orleans from Hurricane Katrina who came to San Antonio by the thousands will be a bellwether for those displaced from other vulnerable megacities like Houston, Texas, or the millions of migrants leaving the equator who cannot be held back by walls.

As I write this book, the Intergovernmental Panel on Climate Change released its terrifying report about a world that is 1.5 degrees to 2.0 degrees warmer.[17] At the time I write this in the summer of 2019, we are currently on a path above 4 degrees Celsius of warming by century's end, and even if we immediately institute all of the Paris climate agreements, we are likely to get 3.2 degrees of warming, which inaugurates the collapse of the planet's ice sheets.[18] So, what we are talking about is an end for many species, many worlds, by the time say my five- and nine-year-old children are my age—in their late thirties—and at the rate we're going, my children may never have grandkids. Meanwhile, immigrant women are detained in tents and raped, screaming children are separated from their mothers, and a political monster who concentrates the worst aspects of this colonial country currently controls the federal government. Human beings pour over the borders by the thousands fleeing from war and the ravages of climate change in Central America. None of this is new. None of these are separate issues. If climate science gives us the profound power of partially foreseeing the coming climate events, perhaps one role of rhetorical scholars is to listen to the multiple voices from previous climate catastrophes and to understand the obligations that privileged positions have to these people. In this way, extreme weather events can become causes for thinking about how to make politics differently and build a capacity for creative ways of surviving through the coming catastrophes and their political struggles.

My desire to write about climate politics in San Antonio, Texas (SATX), comes from a desire to understand how local publics, politics, and environments feed back and co-produce one another in a world of supercharged weather now driven by anthropogenic climate change. Engaging publics marked by difference requires me to interpret specific cultural, material, and rhetorical practices that are not my own and thus to take on the position of an outsider. For outsiders, exposure matters, and my own lived experiences have helped this process of understanding and interpretation that is simultaneously close and distant. In writing this book, I have attempted to follow the lead of San Antonio's artists, scholars, and activists and maintain dialogue with them about my research and writing practices. All of these experiences have furthered my exposure and enriched my understanding of San Antonio's deep and multidimensional sense of belonging. I recognize I can never fully capture the experiences of Mexican American communities, and I recognize that I as a white individual with settler histories in this country have historically benefited from racial oppression. *Whose protection and whose progress?* I am deeply indebted to all those scholars, artists, and activists who have shared their work and their life with me during this book project. I am humbled by the opportunity to converse and work with them. Those people still living who show up in this book know full well I am writing about them, and they have given me permission to quote their words and amplify their voices to scholarly and public audiences. If you find any brilliance here, you should go read them.

All communities and political movements have their outsiders: liberal white professors and el movimiento; Frederick Douglass and the women's liberation movement; and the Anglo cops who decried and denounced the anti-Mexican violence of the Texas Rangers, sometimes risking their lives.[19] Those of us with the privilege of choosing our associations must be careful to not falsely claim membership in communities we do not belong to. I do not have the privilege of belonging to a community of Mexicana/o and Latina/o-origin peoples. But as Phaedra Pezzullo teaches us, ultimately, it is less helpful to decide who is in and who is out—to be fully co-participant or retain a critical distance—than it is to maintain dialogue and conversation. This is a position located between "the tensions of detachment and commitment, objectivity and subjectivity [that] . . . works beyond the wrong assumption that we are not implicated in each other's stories."[20] Indeed, my personal and professional experiences have told me there is much more in common in cross class, race, gender, sexuality relations than there are differences. I recognize, accept, and celebrate those differences. Here I make an honest attempt at putting them into equality as a small gesture toward shared practices and potentially shared politics.

Yet, scholars who are outsiders to some of the communities they study must recognize how the entire context for their scholarship is constituted by inequitable modern/colonial forces at nearly every level—politics, economics, knowledges, gender, and the social, just to name a few. But this is one lesson of border-thinking: As a white scholar at a majority Black and brown institution of higher education I am always already part of the problem. My white male cisgendered body in a location of privilege to research and teach quite literally perpetuates historical and systemic structures of racial injustice and white privilege. Even in the most generous reading, I am complicit in a long legacy of colonial power and state institutions built to sort working- and middle-class students, many of them multiply marginalized, into economic classifications where they can produce value for the wealthy owner class of our contemporary society. It doesn't matter what I research, teach, or write. Just this fact alone is damning. Somehow, I made it while most scholars of color did not; furthermore, they didn't make it partly because people like me did. There's no escaping this fact. There's just an owning of it. And then, there's a decision to be made about what to do with all these privileges.

While such decisions can never be reduced to a recipe of strategies, the responses my rhetorical colleagues have crafted are instructive. From Christa Olson, I have learned to reflect critically on my own racialization and institutional position; to acknowledge the legacy of colonialism and imperialism as constitutive of nearly every aspect of this world; to learn to listen, read, and follow those writers, scholars, performers, and artists who speak from subjugated positions, particularly those who suffer multiple forms of oppression, and especially those from the regions of my sites of study.[21] Linda Alcoff offers an important way forward for rejecting "a general retreat from speaking for" and "a return to an unselfconscious appropriation of the other."[22] Her strategies of analyzing and questioning the impetus to speak, interrogating one's location and context of writing, holding oneself account-able to what one says, and recognizing and examining the effects of one's claims or representations on material-discursive contexts are helpful. As Ol-son notes, neither dominant nor subaltern groups can be fully understood except in relation to one another, and of course there are always internal conflicts within and among groups and subgroups. The stories in this book should remind readers of all the layers of complexity I could not capture, the contested and contestable stories I could not tell, and all the moral ambigu-ity of the silences surrounding whatever wrinkled path you may follow in these pages.[23]

Despite my best attempts to work beyond them, this project is limited by my own ways of living. Those limitations haunt these pages. And they also give them force. Hopefully they generate more stories and counterstories that I cannot possibly tell. I can only claim that I have learned to speak *from* the locations I inhabit in San Antonio, and that the expressions in this book are true to my lived experience. I only ask my honest critics to look to my own academic practices in teaching, service, and research, and in my own personal life, and hold me accountable.

Through the research and writing of this project, I have found speaking *from* certain embodied locations, speaking *for* as an advocate, and speak-ing *with* in solidarity to be practices that foster mutual learning and respect much more effectively than silence. May I continue to learn to speak *from* as I also speak *with*; may I continue to not speak for, but when the moment re-quires it, to speak *for*, as an advocate who lives intentionally and deliberately in San Antonio and with certain San Antonians, and who is invested into certain futures of San Antonio. In writing this rhetorical history of San An-tonio's climate politics, I am an advocate for the perspective that it is largely, but not exclusively, people of color in this city who have led on issues of political equity, and it is largely, but not exclusively, white populations who have refused to listen and follow their lead. This is a structural production. It haunts all of us. I do not claim to speak for Chicana/os, which is impos-sible for me to do. But I do advocate that everyone reading these pages seek a broader and deeper understanding of Chicana/o histories and cultural ex-pressions, particularly because they are leaders in democratic practice, pol-itics, and cultural forms of transnational belonging desperately needed to-day. Having done some of this work myself, I know enough to know that this book is not likely to contribute much to Latina/o rhetorical scholarship than what is already known. I can only claim that in pursuing a story about climate and politics in San Antonio, I have *followed through* in my attempt to

[handwritten margin note: you can allow this information to change how you work]

tell a more complete story because I believe this can tell us something about celebrating diversity in democracy, and indeed truly celebrating America's Indigenous, mestizo/a, and African origins through the climatic changes already here and on the horizon.

In this book, it is generally true that the characters working across communities of difference have more solutions for climate politics than those nursing old grudges. Still, old grudges partially exist because colonial social orders show so little structural change. So, we must consider: What if the specter of climate breakdown is not centered on the limits of scientific epistemologies but rather on another cycle of whiteness triumphing through neocolonialism in a violent and vengeful way? What if climate breakdown perpetuates another wave of colonialism worse than we've ever seen? Worse than the dust bowl. Worse than the genocide of the Indigenous peoples. Worse than slavery. What if the end game of climate change is the worst form of apartheid the world has ever seen? Scholars have to imagine the possibility that climate change could become the final gambit of colonial whiteness on a world stage—an end game of exploitation, where even the planet seeks to destroy the poor and subaltern while the rich and dominant escape through mostly unearned privileges of mobility, invisibility, and transferability. Climate change could be a multigenerational monster for most working people. *Who can be protected in these kinds of worlds? What can progress mean in these worlds?* This is a haunted emergenc(e)/y—like never before, and yet crucially dependent on the scales of human action today, the past will come to haunt the entangled futures of all living creatures who stand to be dominated by the coming catastrophes. We must learn to think with the many worlds of climate and weather events if they are to be anything other than a reanimation of colonial monsters.

CLIMATE POLITICS ON THE BORDER

Introduction

Local Histories and Emergent Futures of Climate Politics

NOT MANY PEOPLE REALIZE THAT South Texas is one of the most climate-aware regions in the United States. According to Yale University's 2017 US climate public opinion map, South Texas stands alone as the only region where publics rank 7–17 percent above the national average of people who think climate change will harm them personally. In Bexar County, which includes the city of San Antonio, more people think climate change will harm them personally than in Marin County, California. In Starr County, on the border between Laredo and Corpus Christi, this figure is 57 percent, the highest in the nation. The reporting on this study suggests that South Texas has "felt the brunt of shifting weather patterns, including rising temperatures, coastal hurricanes and western droughts."[1] Simultaneously, this region has the nation's highest concentrations of Latina/o youth who have embraced climate science to a greater extent than their elders. But climate change and generational change alone do not explain why South Texas publics have internalized the threat of climate change so deeply. This internalization of climate risk suggests public opinion polls are only a glimpse into a larger narrative about climate effects rooted in the histories of this place, its people, and its politics.

At a recent climate justice meeting in San Antonio, Jessica Guerrero, born and raised in San Antonio's Southeast Side, tells a story of how her lived experiences led her to climate justice, particularly around the issue of housing: "I came in direct contact with a lot of different ways and styles of living and you know the one common thing was not economic stability, it was not race, it wasn't any of those things, it was just happiness, right, it was dignity, it was people feeling that they had a place to call home. And it wasn't just about that place where their bed was located, or their kitchen was situated, it was about the whole area, right, that I love, which is the Southeast Side."[2] As Ms. Guerrero spoke about her work to track the effects of gentrification around the San Antonio missions, she lamented that her neighbors saw gentrification in terms of a natural disaster caused by the city. Meanwhile, city officials and San Antonio publics generally seemed to take care of protecting the aquifer and free-tail bats more than displaced people. For Ms. Guerrero, the lack of admitted correlation between numerous city-beautifying projects and housing instability was palpable. She ended her talk by asking her audience: "What is home to you? And what [are] the aspects that make that space that area that feeling home to you and how do we bring you know that knowledge and that conciencia into these spaces of meetings where we need to come out with tangible results. I feel like that's what's missing a lot of the time."[3]

For Ms. Guerrero, a primary source of climate justice is the dignity that comes from the lived experiences of multiple generations of people who create a sense of belonging to a place, a homeland. As a form of knowledge, this conciencia (consciousness) is produced from a long-term relationship among people and places that she was struggling to translate into tangible policies.[4] The continual displacement of her neighbors broke up support networks for people and their place-based practices of care: "We've taken care of the river all these years and we've taken care of each other, and they want to come here and act like none of that matters." In the Q&A after the talk, one young man stated the problem this way: "I feel like we have to name what is happening . . . and we're glazing over the issue of . . . of colonization. And how we are just being . . . people of color are just being displaced over and over again. And I feel like in order to change things we need to say it out loud, colonization, we're still being colonized to this day."[5]

Climate and the contemporary legacies of colonization—this is the subject of *Climate Politics on the Border*, a rhetorical project about the histories and futures of climate and its politics in the colonial multiverse that is San Antonio, Texas. From this location, the national statistics that reveal a deep internalization of climate effects in South Texas are not new; rather, they are the expression of a multigenerational experience of extreme weather and colonial political structures that have crystallized in the opinions of the current generation. Similarly, local battles against contemporary displacement are not new; they are another expression of a long and dignified history of land struggle in San Antonio. Today these struggles are stark reminders of what is at stake in the local politics of climate change—an unprecedented and unescapable cycle of conquest through supercharged climate events.[6] Understood in these ways, San Antonio's politics cannot be dissociated from the layers of colonial histories that have long established power relations in this city—the political dominance of Spanish settlers in the early nineteenth

century; the post–Civil War ethnic Europeans who further established industrial commerce and free market capitalism; the rule of a white and wealthy
Anglo minority during the city's entire modern history; and today's diverse
city council making decisions for this emerging cosmopolis that attracts people from all over the world.[7] Meanwhile, due to the ceaseless extractivism of
modernity, eight of the hottest summers on record in San Antonio have all
been in the last decade;[8] hundred-year storm events (over ten inches of rain
in twenty-four hours) are now twice as likely;[9] and nearby the city of Houston has experienced three five-hundred-year flooding events in the last three
years.[10] The pace, scale, and severity of such events are unprecedented, but
the relationships among climate, people, and politics are not new; rather,
these relationships have place-based histories rooted in the political power
of state settler colonial practices in the United States and beyond. What we
know from these histories is that extreme climate and weather events tend
to reinforce and even strengthen existing power structures. And yet because
these events make long-standing structures of domination radically visible,
there is immense value to studying the fissures and folds of local histories of
climate politics as auguries for future manifestations of adaptation, inhabitation, coalition, and resilience.

Historically, Texas is one of the most vulnerable places for naturally occurring climate and weather-related disasters, but to live in Texas today at
the end of the second decade of the twenty-first century is to live at the prec-
ipice of a climate shift where no one yet knows what the ultimate consequences will be. That uncertainty is not so much about science as it is about
the open question of how quickly humans will respond to scientific, human,
and earthly demands for fewer carbon emissions and fewer colonial relations of systemic dominance. Given our current trajectory above 4 degrees
Celsius of warming,[11] we do know that formerly outlier extreme weather patterns will shift and become new norms, and we know that vulnerable southern states like Texas will face the most extreme climate effects first.[12] In this
way, Texas stands as a kind of bellwether for the effects of human-induced
climate events that will amplify existing extremes across every region of the
United States. But climate change is not just about the dramas of extreme
weather events—the fires, droughts, and hurricanes we all know so well by
now. What we call climate is an encompassing system that so intimately gov-
erns every aspect of life on this planet that no matter where you live there is
no escaping its changes. Extreme weather events are only some of the most
dramatic expressions of these changes. Their frequency and intensity, and
the dire warnings that multiple catastrophic weather events will occur in
certain regions as early as 2040,[13] have all made it perfectly clear that the

ence of a supercharged climate is no distant abstraction. This book is
on the premise that the transformations on the horizon for the United
States are converging now in Texas: climatic change, demographic change,
and an entrenched political system where the gap between people and political representatives is wider than it has been in generations. For better and
worse, one future of adapting to climate change in the United States is happening now in Texas.[14]

Through detailed archival research, interviews, fieldwork, and theoretically informed critical analysis, I contextualize and craft an account of San
Antonio's climate politics to demonstrate how rhetorical theory and practice are poised to respond to these changes. Rhetoric is an art of communication, or more specifically, an art of situated public symbolic practice—it
is radically contextual (situated), it addresses things that are common and
shared (public, though the definition of public is always contested), and it
is a material-semiotic symbolic practice (an exchange and conveyance of
meaning through embodied practice always *done with* more-than-human
environs and creatures).[15] As an art, rhetoric is both an object of study and
a way to study, a hermeneutic and an inventional heuristic that can diag-
nose and guide democratic responses to local climatic disruptions. Yet doing
rhetorical theory and practice in San Antonio demands an epistemic shift
away from rhetoric's European histories and toward a privileging of rhetoric's other possibilities situated within specific places and power dynamics
unique to the Americas. To demonstrate rhetoric's potential for using local
climate and weather events as sites for de/colonial practice, I develop a theoretical framework invested in ecological[16] and decolonial[17] approaches to rhetorical analysis. Through this framework, I then interrogate how rhetorical
practices around extreme weather events have been enacted by a variety of
local publics over time to attempt political change. Thus, this book engages
the histories and futures of climate politics in San Antonio for two reasons:
first, to enrich contemporary discussions about rhetoric, ecology, and de/co-
lonial praxis; and second, to articulate a few discourses and practices for relevant climatic response-abilities.[18]

Rhetorical scholars have yet to offer an orientation to an ecological and
de/colonial praxis of local climate politics, but postcolonial and decolonial
science studies scholars have begun to consider the pluriversal as a viable
response to the modern/colonial extractivism of the Anthropocene.[19] Pluriversal is a political concept first developed by the Zapatistas to describe a
world in which many worlds coexist through and against the power differentials of colonial social orders. As such, the pluriversal is first a decolonial
political vision of a world of many worlds. In Marisol de la Cadena and Mario Blaser's refiguring of the concept, the pluriversal is a reconsideration of

"the material-semiotic grammar of *the relation* among worlds" that is useful for "conceiving ecologies of practice across heterogeneous(ly) entangled worlds."[20] Reconsidering relations among worlds must be practiced in specific contexts, but in general ecological and decolonial scholars are deeply invested in transitions away from the one-world modern/colonial development model that has brought many worlds to the brink. At present many argue it is easier to imagine the end of the world rather than the end of modernity/coloniality—how systems of social hierarchies, knowledges, and cultures that privilege Eurocentrism have evolved from colonial contact.[21] But to commit to the study of the relation among worlds is also to stay with the trouble and ask critical questions about what kinds of climatic response-abilities can come to relevance in the context of climatic breakdown.[22] Such a material and relational approach to the pluriversal as a speculative form of response-ability is particularly suitable for ecological approaches to rhetorical theory and practice. Thus, for the purposes of this book, pluriversal rhetorics are those that study or practice rhetoric as an ecological (a material and relational) system that functions across entangled worlds and against the power differentials of coloniality. When pluriversal rhetorics engage with local climate politics, they yield heterogeneously relevant response-abilities to climatic breakdown.

Jargon-Fill [margin annotation]

"pluri-versal rhetorics" as rhet. type specific to EJ [margin annotation]

Thus, pluriversal rhetorics are specifically designed to provide answers to the overarching questions driving this book: First, how do climate and weather events function rhetorically and politically over the long term across situated local publics characterized by difference? And second, how can this rhetorical knowledge help local and national politics better prepare for climate-changed futures? In answering these questions, *Climate Politics on the Border* attempts to make three distinct contributions to rhetorical theory and practice: first, this project is an inaugural attempt to address local climate politics from the standpoint of rhetorical analysis; second, it offers a rhetorical methodology—pluriversal rhetorics—for the analysis of situated public rhetorics across entangled worlds marked by the colonial wound; and third, it tells a local story about how commitments to deliberation and decision-making through difference and climatic events constitutes a slow but steady arc toward political inclusion and climatic response-ability that will likely be entirely outpaced by the scale, severity, and scope of our contemporary climate crisis. To make these contributions to rhetorical scholarship, this project is primarily informed by ecological and environmental rhetorics from rhetoric of science, technology, and medicine (RSTM) and its associates in science and technology studies (STS), and decolonial praxis and its associates in border rhetorics, Latinx rhetorics, and environmental justice rhetorics. I undertake this project for two primary reasons: first, blish a conversation among these various subfields, and second, to :cological/environmental rhetorics broadly, and RSTM specifically, to ueep n its investment in a de/colonial praxis of science and its politics by addressing the power differentials of coloniality.

As an archetype of the Sunbelt American city, San Antonio is both a unique and instructive site for such a theoretical orientation—it has a long political history of colonial relations through extreme climate/weather events that has largely gone unaddressed by scholarship, yet, its rhetorical histories and futures offer rich and textured examples of how publics, places, and power relations are negotiated through the experience of colonial and climatic extremes. Because of its multiple generational knowledge of adapting to extreme weather through the structural inequalities of colonial contact, San Antonio offers rhetorical studies a unique borderlands vision of climate politics. At the same time, understood rhetorically, San Antonio is a metonym for every city seeking to become resilient in a world bracing for the effects of a climate breakdown. The rest of this introduction outlines how pluriversal rhetorics can approach climate and its politics to resist, erode, and alter everyday colonial social orders.

PLURIVERSAL RHETORICS FOR CLIMATE POLITICS

Etymologically, climate shares a close relationship with the materiality of rhetoric and politics, particularly through commonplace notions of what _prevails_—a practice of enduring, widespread, and persuasive power, as in a prevailing climate or a prevailing public opinion.[23] Scientifically, climate is defined as a region with a prevailing weather pattern, or the average weather conditions in a region over a roughly thirty-year period. More rhetorically, climate is defined as the prevailing trend of public opinion or public life, codified by the phrase "political climate." Thus, at its core, the concept of climate is profoundly regional and rhetorical—it connects place-based material conditions (environments) and public material-semiotics (meaning-making) through the situated norms of creating worlds together (politics). In this sense, climate is the mutual shaping and coexistence of environments and politics together, or as I will argue throughout, climate is a prevailing political embodiment. As an art of situated public communicative practice, rhetoric provides an inventional and analytic approach for studies of climate as a science and as a politics. In this view, rhetoric is not just a presidential speech, a testimony, crafty language manipulation, or bullshit; rhetoric is a material and semiotic performance that constitutes public identities and shapes political communities by moving people to collective beliefs and actions.[24] As such, rhetoric is central to the politics of democracies, and climate gives rhetoric material force.

The associations between climate and rhetoric are typically experienced as an everyday discourse of praise and blame about weather events that humans often deny has any human dimensions. Such weather events are "natural" disasters, amoral chance occurrences, and acts of God.[25] While partially true, such explanations obscure the human, social, and political forces that continually place working people and people of color in precarious living conditions. In fact, "natural" disasters are historically produced through human choices in relation to climate and weather, and thus they are legitimate subjects for social, historical, and rhetorical study. This scholarship points out that "the concept of 'natural' disaster developed when those in power in disaster-stricken cities sought to normalize calamity in their quest to restore order, that is, to restore property values and the economy to their upward trajectory."[26] To call such disasters "natural" thus operates as a rhetorical maneuver of conquest that normalizes precarity, or those "politically induced conditions in which certain populations suffer from failing social and economic networks of support and become differentially exposed to injury, violence, and death."[27] Similarly, to claim that a city's inhabitants should "adapt" or be "resilient" to supercharged climate and weather events is also a form of conquest when such rhetorical practices reinforce structural inequities and/or make them worse.[28] Against these narratives, this book begins from the understanding that when politics is placed in its proper colonial context there is very little that is "natural" about the effects of extreme climate and weather events that precariously placed people are supposed to "adapt" to. Of course, this is only more horrifically true on hothouse Earth where the world has pumped more than half of all the fossil fuel carbon pollution into the atmosphere since the first episode of *Seinfeld* aired in 1989.[29]

With this understanding, this book offers a rhetorical history about how the material forces of climate have shaped San Antonio's city politics, and how politics have materialized understandings of climate—or what I call climate politics. San Antonio is a city with a long, multitudinous history of colonial contact, extreme weather, and public rhetorics of equity due to its location near the US-Mexico border and its status as an archetypical Mexican American city that city officials often tout as a place where "diversity is done right." To complicate these narratives, I engage pluriversal rhetorics as a critical corrective to a city that is also ranked among the nation's highest poverty and most economically segregated cities. Pluriversal rhetorics are deeply influenced by border thinking, which arises from the point of colonial and cultural contact and entails the disruption of dichotomies by bringing subalternized discourses and practices into articulation with dominant logics,[30] and border rhetoric is the expression of border thinking.[31] In bringing subaltern discourses and practices into articulation with dominant logics through climate politics, my purpose is not to situate colonial difference as an object of study, but rather to situate colonial difference as a rhetorical practice with potential for new relations in adaptation, resilience, and placed-based belonging.[32] For certain scholars with a primary orientation toward ecological/environmental rhetorics, bringing subaltern discourses and practices to bear on climate politics means questioning everyday manifestations of coloniality in order to reconsider relations among worlds.

Pluriversal rhetorics for climate politics are therefore not just a matter of science and hegemonic politics but a matter of how multiple generations of people continually (re)figure climate as meaningful, persuasive, and political through their everyday lives.[33] A generation of climate science tells us that solving the problems associated with climate change means simultaneously solving the problems with democratic politics;[34] yet, while climate science continues to look to history for auguries of future climates (i.e., paleoclimatology), not enough work has been done to examine histories of climate as political forces. The consequences of this epistemic neglect are significant. Consider: for as long as scientists have studied planet Earth with an identifiable and generally knowable global climate system that fosters the conditions for life on this planet, regional and local knowledges of climate have existed for a much longer period of time. In drawing attention to cultural understandings of climate, cultural geographer Mike Hulme redefines climate as *weather ways*—"an idea which mediates the sensory experience of ephemeral weather and the cultural ways of living which humans have developed to accommodate this experience."[35] Importantly, climate as weather ways is always placed. As a science, climate describes what weather is; as a politics, climate describes what weather does politically.[36] Thus, there is no single thing called climate change. There is only a pluriverse of climatic discourses and practices, and this pluriverse must be mapped locally and regionally (in addition to nationally and globally) if one is to understand the full implications of climate change as political change. The same is true of the effects of climate change—they will not be singular but rather a pluriversal manifestation of colonial/modern monsters that manifest through everyday experiences and worldviews.

As a decolonial concept that includes consideration of more-than-human, the pluriversal has dramatic implications for rhetorical theory and practice that by definition are not exclusive to Indigenous rhetorics. As Ellen Cushman has written recently, the effort to move toward pluriversal possibilities must involve everyone who is situated within the colonial matrix of power, and it begins with an epistemic de-linking that seeks options to the power differentials of coloniality/modernity. Importantly, this is not a project based on cultural relativism, nor one that seeks alternative rhetorics, but rather one that dwells in entanglement and the borders of the colonial matrix of power. This is another way of stating the value of Darrel Wanzer-Serrano's inclusive project for de/coloniality, which seeks to pluralize rhetorical praxis rather than authenticate or universalize alternative approaches to rhetoric.[37] Engaging in pluriversal projects, therefore, is never just a matter of cultural appropriation, a phrase that itself assumes a static, original, concept of culture rather than a fluid, radically situated, and creative process of ongoing exchange. Still, the power differentials of coloniality are not metaphorical, and pluriversal projects must also account for how they materially benefit marginalized peoples and communities, especially Indigenous communities.[38] As I use the term in this project, pluriversal rhetorics are particularly suitable for interrogating the intersections of ecological and de/colonial theory. The promise of such a project is that through rhetoric it helps create a space for ecology to come to terms with the legacy of colonialism and imperialism, and for border thinking to come to terms with ecology and a world of many sciences.[39] White settler scholars such as myself have much work to do in this regard since people in my own gender/race/sexuality categories created and sustained coloniality in the first place. Therefore, it is also up to people like me to help dismantle it and seek to foster a world where many worlds coexist without the extreme power differentials of our contemporary moment. In these ways and for these many worlds, ecology, de/coloniality, and pluriversality are everyone's responsibility.[40]

Of course, such projects do not begin from scratch but rather from overlapping conversation points across scholarly areas that may yet have a history and future worth reknotting, reworking, and rethinking for each particular situation. These overlaps do not mean ecology and de/coloniality do not have different histories or tensions; their relationship in this project might best be characterized by the notion of an un/commons—a negotiated coming together of heterogenous worlds and practices that strive to be what they are that is also not without others.[41] This book submits that making kin in this way is not just the responsibility of de/colonial scholars in rhetorical studies but also the responsibility of ecological, environmental, and science studies rhetorical scholars who must work to make room for diverse and subversive understandings of sciences from below[42] and from elsewhere/otherwise.[43] I believe one way forward on this path is for rhetorical science studies scholars to withhold big claims to decoloniality and instead examine the everyday effects of coloniality in their technoscientific sites of study. For readers interested in a rhetorical approach to conversations on coloniality and climate politics, read part 1. There I describe in detail how ecological rhetorics and border thinking can serve as theoretical bridges toward pluriversal rhetorics. For nonscholarly readers, I recommend you dive right into the San Antonio narratives beginning with part 2.

A RHETORICAL HISTORIOGRAPHY FOR PLURIVERSAL CLIMATE POLITICS

As a rhetorical analysis of local climate politics in San Antonio, this book's focal points are those public discourses and practices around climate and weather events that seek sociopolitical transitions, transformations, and/ or intransigence. For example, when someone blames the effects of flash floods on God's displeasure of human behavior, or on neglect from city politicians, those are climate politics because they testify to a sociopolitical order and seek some kind of differential action based on the premise of the argument (i.e., floods are God's wrath, seek salvation; floods are social disasters, fight for political transformation). More specifically, this book uses rhetorical analysis to trace a history of antecedent climate justice politics and to rethink contemporary notions of city-level climate adaptation. Climate adaptation is often compared and contrasted with climate mitigation, which seeks to stop the flow of carbon pollution into the atmosphere, popularized by slogans of "zero emissions" or "leave it in the ground." While the fight to mitigate the flow of carbon pollution into the atmosphere dates back to the 1980s, climate adaptation is a relatively recent phenomenon most visibly discussed in the second assessment report from the Intergovernmental Panel on Climate Change (IPCC), published in 1995. There, climate adaptation is defined as the "adjustments to natural or human systems in response to actual or expected climate change, including increases in the frequency or severity of weather-related disasters."[44] Adjustments to coupled natural and human systems means climate adaptation can be incremental (maintains systems integrity) or transformational (systems change), and of course is highly variable by location. The adaptive capacity for any given region is closely tied to social and economic development, and thus adaptation raises issues of equity and political power since adaptation tends to work best for those who can afford it. Indeed, the ease with which adaptation discourses and practices are co-opted by existing power structures is what led the trade union movement to develop the notion of a *just transition* that centers concerns of equity and justice as economies transition away from fossil fuels. A just transition has been widely applied in key moments like the Paris Agreement and in the Green New Deal, and this project is specifically concerned with its implementation in San Antonio.[45]

Today many communities across the United States are realizing what adapting to climate and weather extremes looks like in practice—the wealthy and the elite escape and then return to take advantage of disasters by strengthening their social positions, often at the expense of marginalized groups.[46] The lessons from Hurricane Katrina, Superstorm Sandy, Hurricane Harvey, and others are that "natural" disasters further concentrate wealth and uneven recovery, and they perpetuate cycles of conquest ruled by a colonial mentality of "in devastation there is opportunity." In the words of climate advocate David Roberts, the call to be more "resilient" during these events means that for many, public rhetorics of climate adaptation are a cruel joke—a condescending call for working people to "move to higher ground" without providing the available means of doing so.[47]

My own examination of the political effects of extreme weather over time in San Antonio suggests that the discourses and practices of climate adaptation largely accomplish four things: they often reinforce and strengthen dominant forms of social power; they set stages for eulogies of loss and survival, sometimes heroically so; as tropes of power, they create sites of political struggle that can be used differentially; and the extreme weather events they are based on function as analogical verifications of sociopolitical conditions. While one's ability to adapt and be resilient is often a factor of wealth, the political potential of these concepts partly lies in a sense of belonging to places and neighborhoods with strong social networks established to create plans, provide early warnings, and use public and commercial spaces for systems of aid and support. As Ms. Guerrero notes in her work on housing justice, adaptation is also about the dignity and happiness that comes through the creation of a home place with a deep sense of belonging among people, places, and more-than-human communities. Regardless of monetary wealth, studies consistently show that communities with a strong social infrastructure are better prepared for adaptation to climate forces.[48] So, the questions about climate adaptation for each location become, what kinds of sociopolitical transitions and transformations are possible through occurring or anticipated extreme weather events? And relatedly, how can one use these events as a cause for creating a more widely shared social and economic equity to thrive through climate change and not simply survive it? To Ms. Guerrero's point, how can one live through these changes with dignity, happiness, and a strengthening of local attachments?

One way to answer these questions is to rhetorically examine placed-based histories of modernity/coloniality and their relationships to past and current climate and weather events as issues of climate justice. At their root, the dramatic structural inequalities found in most US cities today are the consequence of evolving colonial systems, and thus, truly transformative notions of climate adaptation and just transitions are best framed through a colonial systems analysis. This is why climate adaptation needs to be properly historicized and politicized through border thinking—because border thinking creates a third space that challenges systemic colonial orders of race, class, gender, sexuality, economy, and more. From a borderlands perspective, the work of housing justice is an issue of climate adaptation, for example, because housing captures those differential ways of creating a social infrastructure so important to adaptation. A deep understanding of local histories of arrangements of public goods and services for each neighborhood matters for notions of a just transition and adaptation to the local effects of climate change. Thus, one major goal of this book is to provide pluriversal pathways for understanding the rhetorical relationships between climate and politics in the process of place-making. In this way, this project attempts to bring rhetoric to the forefront of our contemporary political conversations of climate change as political change,[49] and it does this through the analysis of a multigenerational site of border thinking about climate adaptation—San Antonio, Texas.

Admittedly, one does not tend to think about Texas first when one considers a just form of climate adaptation. Quite the opposite. Today, Texas stands as a paradox of climate politics: it is the world's seventh-largest emitter of carbon pollution, yet it is also the leader in wind energy in the United States; it has a long history of colonial contact and racial segregation, yet cities like San Antonio are often pointed to as models of democratic inclusion; Texas is already vulnerable to extreme weather that will get worse, yet more people are moving to Texas cities than ever before in the history of the state; and cities and increasingly the suburbs in Texas are politically blue (and therefore drafting climate action and adaptation plans), while there is more infrastructure for oil and gas in Texas today than at any point in its history.[50] In short, for climate politics, Texas has got it all, y'all.

In order to properly understand how these historical, political, and environmental formations came to be, this project takes both a field-based and pan-historiographic approach wherein I examine moments of significant interaction among climate, cultures, and politics across one hundred years of San Antonio's modern/colonial history.[51] Because subjectivities and political practices take place over the long term, pan-historiographic approaches to climate politics can help clarify moments of resistance and transformation that otherwise may go unnoticed in a narrower framework.[52] In turn, pan-historiographic perspectives provide clues to the contemporary global/local politics of climate risks at a variety of scales, including the scale of everyday practice and institutional decision-making. Thus, a pan-historiographic rhetorical analysis affords a textured understanding of rhetorical practices over time in a single place, which should yield both critical placed-based interventions and contributions to scholarship.

One of those contributions that this book makes is to interrogate the c nections between pan-historiography and rhetorical field methods for the purpose of a placed-based and engaged critical rhetoric. A number of recent rhetorical works have demonstrated the value of field-based rhetorical methods, particularly for ecological engagement in place-based studies.[53] Because this book asks questions about place, power, and politics over a century, field methods help capture local and vernacular material-discursive practices that would likely go unnoticed with a strict focus on texts. Field methods also allow for an analysis of material and cultural spaces where rhetorical critics can engage in analyses of topographies, species assemblages, urban designs, watersheds, performances, and the feelings/affects of embodiment that a strict textual approach to rhetorical criticism cannot capture.[54] Ultimately, the choice of method should be guided by a researcher's questions and goals, and because this book oscillates through dimensions of history, present and future, it finds both historical-archival and fieldwork conducive for its questions about the relationships among climate, publics, and politics. Indeed, it is when historical narratives are reinforced by contemporary observations, informal interviews, field experiences, and vice versa, that a version of evidential triangulation becomes possible for claims about the continuing legacies of climate politics. It is this mix of de/colonial rhetorical historiography and field-based ecological engagement that offers one methodological approach for pluriversal rhetorics.

Through this methodological orientation, *Climate Politics on the Border* tells a story about how actual and anticipated extreme weather events like flash flooding catalyze a city's development struggles, particularly marked by four key moments in San Antonio—the decision to build Olmos Dam after the 1921 flood; expansive suburbanization in the mid-1970s; urban core redevelopment in the contemporary moment; and the future development visions of city-level climate action and adaptation plans. I chose these moments in particular because they illustrate the connections among climate/ weather events, situated public rhetorics, and political decision-making. Each one of these moments carries diverse meanings of climate into its public rhetorics, and yet which of them count as meaningful and persuasive for citywide policies is the specific rhetorical locus of this study. Together these events tell a story about why adapting to climate and its changes is always about the capacity to create a kind of publicness, a prevailing political embodiment, and thus a cosmic vision of the human and more-than-human place-making in the city. In this sense, this book is a story about the prevailing embodiments of publicness in San Antonio over the course of modernity/coloniality. This story reveals that while Mexico-origin people have gained substantial political inclusion and power over the last half century, in a city with a population that is 65 percent Latinx, San Antonio could do much more to shatter its illusions of inclusion.[55] As a story about public embodiment, San Antonio reminds all Americans in the United States that Mexicana/o and Chicana/o histories continue to leave indelible marks on United States history, and they demand much more public recognition. Places like San Antonio remind us that the Indigenous populations of the United States, the rise and fall of the Aztec Empire centered in Tenochtitlán, the story of Spanish contact and subsequent conquest, and the emergence of New Spain, then Mexico, and then the entire Southwest and most of the western United States, is as "American" as Plymouth Rock. Thus, to study climate politics in San Antonio is to study what author John Phillip Santos calls an *ambiente Mexicano*—a uniquely rich Indigenous and mestizo American legacy with a long-standing relationship to its southwestern climate.[56]

In order to maintain a dynamic among contemporary debates in climate science and regional histories of climate, I begin each chapter with a common matter of concern from contemporary climate science, often through the work of Katharine Hayhoe, who more than most has focused her work on down-scaling climate science to regional and local levels, especially in Texas. In each chapter, I show how the contemporary quality of life problems at the thresholds of contemporary climate science are nothing new. Problems such as resilience to extreme weather events, social and material infrastructure for climate-proof neighborhoods, and sustainable development projects are in fact quality of life problems that have confronted modernity/coloniality since its inception. Each chapter then maps how modernity/coloniality in San Antonio dealt with the issues identified by contemporary climate science at the city level, and the final chapter reflects on a city-level climate action and adaptation plan for a generation that stands to experience an unprecedented world of supercharged climate and weather events. Yet because colonialism is constitutive of every aspect of the modern political world, each chapter also traces dominant and marginalized situated public material-semiotics seeking to affect political change through climate and weather events. Therefore, this project and its capacity to be sensitive to forms of oppression and to forms of resistance is deeply indebted to a long history of San Antonio's community of scholars, writers, artists, and activists, especially those who have engaged with me throughout the research and writing of this book.

Once the contemporary concern of climate politics is addressed, the parameters of each chapter are defined by a specific climatic event in San Antonio—often a flood or the invocation of flooding in order to mobilize politics around issues of security, energy, and public space. Within each chapter, I map the various scientific actors with significant participation in the event—atmospheric scientists, hydrologists, geologists, landscape architects, engineers, and so on. The various scientific actors that appear in each chapter codify a scientific practice of climate politics that I use to examine its effects with an eye toward the types of subjectivities most affected, and what types of civic interventions may have been, and perhaps still are, possible. Each chapter also has its share of public and political actors who participate significantly in each event of climate politics—flash floods, city politicians, environmentalists, Chicana/o activists, community organizers, clergy members, limestone escarpments, journalists, and more. These actors activate specific practices of climate politics that range from culturally specific climate knowledges and practices to highly personal ones, to overtly political ones, and to hydrogeographic ones. These pluriversal climate practices emerge, coalesce, diverge, and conflict in radically contextual manners. The public discourses and practices mobilized by such events constitute the sites of my rhetorical analysis at the microlevel. Because these discourses and practices are instantiated in particular genre ecologies, each event is also studied through multiple archives of genres or text types—scientific reports, meeting notes, news articles, editorials, press releases, pictures, board meetings, design plans, and more[57]—in addition to lived experiences and fieldwork with relevant actors where applicable.[58] Archival work and fieldwork capture the mesolevel of human/nonhuman activities as markers of place-based expressions and thus are an important analytic for emplaced publics marked by difference of all kinds. At the macrolevel of theory, I use pluriversal rhetoric to critically interpret each event of climate politics and discuss its implications for San Antonio and beyond.

As a form of preview for these rhetorical histories and futures of climate politics, let me make the main points of each chapter explicit. As I do, I encourage readers to consider that histories and futures are fluid, and any one moment will carry significance for another moment, but all moments carry orientations toward place-based and bioregional senses of belonging. Thus, while the book is presented chronologically, any one of these moments is indicative of a deep sense of time that folds histories, presents, and futures into pluriversal dimensions where any given story may come to relevance at any given moment. Part 1 contains a single theoretical chapter that is written directly to an interdisciplinary group of scholars from rhetorical studies and science studies who engage in decolonial and/or ecological praxis. The chapter details the intellectual genealogy of pluriversal rhetorics and frames the nonmodern sources of the pluriversal as a joint kinship among ecological and de/colonial rhetorics. Through an examination of rhetorical science studies, engaged ecological rhetorics, and de/colonial praxis via critical border thinking, I offer pluriversal rhetorics as a context-dependent methodology for tracing the entangled and emplaced local histories and futures of climate politics. Nonscholarly readers will likely find it most productive to skip part 1 and dive directly into the San Antonio narratives in part 2.

Part 2 contains three narratives from San Antonio that begin in the Texas modern period and end in our contemporary moment. Chapter 2 offers a pluriversal rhetorical understanding of resilience as it is experienced through extreme weather events. Through detailed archival work surrounding this event, this chapter demonstrates how a historic flood reified colonial structures of segregation and anti-Mexican violence that were also resisted by Mexicana/o public rhetorics of shared sacrifice and technoscientific policies for equitable investments in infrastructure. The ensuing political decision to build Olmos Dam in an area that only protected wealthy Anglos concretized these segregationist discourses, practices, and subjectivities for generations, a legacy that remains to this day. A public reckoning of this legacy, I argue, requires a political practice of shared sacrifice as a necessary fold in resilience to keep this topic appropriately political by thinking in the presence of those most affected by the events and ensuing decisions. Ecologizing these effects of modernity/coloniality must still address the ways in which both science and public rhetorics were distorted to create segregationist landscapes. Ecologizing these histories would also mean openness to the un/commons of science in the service of pluriversal politics.

Chapter 3 offers a rhetorical understanding of divergence and diplomacy to understand materialities as necessary conditions for coalitional politics. Through a rhetorical reading of Isabelle Stengers's and Marisol de la Cadena's ecologies of practice as a form of critical border thinking, I detail how de/colonial and ecological rhetorics contain a shared emphasis on situated praxis for the purpose of pluriversality. To demonstrate I offer a case study of a political coalition between a local environmental group seeking water protection and a second-wave Chicana/o organization responding to violent flash flooding. Through my rhetorical analysis, I argue this coalition could only emerge with and through their specific material conditions that when articulated to broader political forces helped transform San Antonio's city politics toward pluralism. Divergence and diplomacy are viable theoretical concepts for thinking through the material conditions of coalitional politics at level of praxis—an increasingly needed area of rhetorical theory and practice under climate breakdown and the coalitions already responding to it.

Chapter 4 is based on my local fieldwork in San Antonio, and it examines the design and deliberations around a local downtown watershed restoration and redevelopment project to demonstrate rhetorical practices of place-keeping and inhabitation for pluriversal notions of adaptation. The chapter shows how specters of extreme flooding are used to redesign urban spaces, create consumer markets, attract global capital investments, and ecologically "restore" watersheds. Yet, as I show through the deliberative process of this space, these practices use ecology to reify a white spatial imaginary as a central aspect of restoration and development. Only through local government structures creating a protected enclave of Chicana/o artists, designers, and activists did the project resist these public forces and foster some sense of pluriversal adaptation through ecological and de/colonial practices of place-keeping and inhabitation. Yet if these local struggles are indicative of what climate adaptation means today, such incremental politics brings the limits of city-level climate politics into full view, especially given the dire warnings that contemporary politics have severely underestimated the scale, severity, and scope of coming climate catastrophes.

In this context, part 3 starts with chapter 5, which reads San Antonio's Climate Action and Adaptation Plan (CAAP) in light of pluriversal rhetorics and contemporary climate science. Rather than look to the development of the city-level plan, this chapter uses the plan to ground speculations on futures of pluriversal rhetorics and climate politics. Taking a critical view of climate equity, I show how under the auspices of city-state-capital control, equity as fairness all too quickly becomes equity as raised property values. This points to the difficulties of regulating the translation of equity and the need to rethink equity in terms otherwise and elsewhere outside of state-colonial capital control. The rest of the chapter uses the CAAP to speculate on what is coming for intersectional approaches to climate politics—mass migrations and the force of climate, critical rhetorics of everyday energy politics, and creative capacity of disturbance rhetorics when climate breakdown upends any sense of normal. Rhetorical studies will be challenged in unprecedented ways by these happenings, yet the lessons from each chapter provide some guidance toward what a shared good life might mean under civilizational collapse. The coda then offers examples of ongoing pluriversal rhetorical projects happening now and on the horizon. My aim is to show how through pluriversal rhetorics, democracy within diversity, and climate adaptation without assimilation may yet be possible through the coming climate politics.

PART I
Theorizing Pluriversal Rhetorics

CHAPTER ONE

Rhetoric, Ecology, and Pluriversality

This is what articulation does; it is always a non-innocent, contestable practice; the partners are never set once and for all. There is no ventriloquism here. Articulation is work, and it may fail. All the people who care, cognitively, emotionally, and politically, must articulate their position in a field constrained by a new collective entity, made up of indigenous people and other human and unhuman actors. Commitment and engagement, not their invalidation, in an emerging collective are the conditions of joining knowledge-producing and world-building practices. This is situated knowledge in the New World; it builds on common places, and it takes unexpected turns.

DONNA HARAWAY, *The Promise of Monsters*

If a pluriverse is not a world of independent units (as is the case with cultural relativism) but a world entangled through and by the colonial matrix of power, then a way of thinking and understanding that dwells in the interstices of the entanglement, at its borders, is needed.

BERND REITER, *Constructing the Pluriverse*

MY GOAL IN THIS CHAPTER on rhetorical theory is to argue for a broad investment into the intersections of ecological and decolonial praxis and to demonstrate how these investments can be productively approached through pluriversality. The pluriversal is a Zapatista concept and a decolonial political vision of a world in which many worlds coexist.[1] As Bernd Reiter notes in the second epigraph to this chapter, the pluriversal is closely connected to what de/colonial scholars call the colonial matrix of power, or, for short, coloniality. As defined by Aníbal Quijano, coloniality describes how systems of social hierarchies, knowledges, and cultures that privilege Eurocentrism have evolved from colonial contact.[2] The evolution from colonialism to coloniality helps explain how Euro-American and patriarchal social orders are perpetuated long after colonial contact, perhaps most visibly in majority-minority places like San Antonio. The pluriversal acknowledges that all living things are entangled through and by coloniality, and it offers a way of thinking and practicing from the interstices, wounds, and/or borders of ongoing colonial contact. As recently proposed by science studies scholars Marisol de la Cadena and Mario Blaser, the pluriversal is also a response to the Anthropocene as a reconsideration of "the material-semiotic grammar of *the relation* among worlds" that is useful for "conceiving ecologies of practice across heterogeneous(ly) entangled worlds."[3] In reconsidering relations among worlds, the question of how to relate to modernity/coloniality is still an open one and best answered in specific contexts. Exactly how rhetorical studies at large, and particularly rhetorical science studies, can contribute to these reconsiderations of the relations among worlds in order to compose the pluriversal remains a question well worth addressing.

Thus, the overarching goal of this book is to use ecological and de/colonial theories to examine the composition of the pluriversal through a rhetorical study of climate politics in San Antonio. Ecology is a European term with roots in the Greek for home (*oikos*), and it is not an inherently decolonial concept. However, its decolonial potentialities can be highlighted by its history as a metaphorical concept and an actual science that carefully considers relations among worlds—a viable analytic for pluriversality. In their use of Stengers's ecology of practices, de la Cadena and Blaser's work makes it clear that ecological theories have much to contribute to the pluriversal, but understanding rhetoric's relationship to philosophers like Stengers requires further examination in order to understand how rhetorical practices compose these ecologies as acts of creation that resist one-world extractivism and articulate speculative visions of the future. As I noted in the introduction, bringing rhetoric to these conversations partially means emphasizing the situatedness and publicness of the pluriversal, and indeed, this project is invested in those rhetorical theories and practices that address climate and weather events as a relational means of moving publics toward collective beliefs, identifications, and actions.[4] Such praxis demands a scholarly fusion that can help rhetorical scholars trace multiple, divergent discourses and practices surrounding the local politics of climate and weather events in an era partially defined by ecological rhetorics and politics.[5] Yet because ecological rhetorics are connected in a power differential based on logics of coloniality, pluriversal rhetorics are also deeply informed by de/colonial theories, especially from border rhetorics that arise from the point of colonial and cultural contact and entail the disruption of dichotomies by bringing subalternized discourses and practices into articulation with dominant logics.[6] In conceptualizing local climate politics as inducements to de/colonial praxis, pluriversal rhetorics can help address the structures of coloniality and constellate worlds capable of response-abilities to climate breakdown. As I argue here, the pluriversal is thus a rhetorical analytic and heuristic for local histories and futures of climate justice.[7]

The remaining pages of this chapter unfold the contours of this book's rhetorical orientations through the key concepts of the book's main title: climate, politics, and borders. First, I address the notion of climate through ecological rhetorics and rhetorical science studies and argue for a deep investment into the intersections of ecological and de/colonial theories, partially through notions of pluriversal rhetorics as climatic response-abilities. Second, I address how rhetorical techniques for publics, place, and politics can be thought through ecological and de/colonial theories and their overlaps. Finally, I discuss how pluriversal rhetorics are informed by border thinking and practice in order to invest rhetorical analyses of climate politics with the necessary critical analytics of de/coloniality. Importantly, this also allows border thinking and practice to potentially come to terms with ecology and the more-than-human world as decolonial propositions. While each section is distinct, I thread terms across sections in order to demonstrate the value of dwelling at the intersections, borders, and wounds of these subfields. Then, each chapter that follows takes the terms introduced here and explores them in specific moments of climate and its politics over one hundred years of modern/colonial history in San Antonio, Texas. Throughout I hope to show how rhetoric can be productively brought to the forefront of our contemporary conversations of climate change as political change.

THE POLITICAL CLIMATE OF RHETORICAL SCIENCE STUDIES

As an art of communication, rhetoric is closely associated with democratic practice and politics, and ecologically inflected rhetorical arts have a rich tradition of scholarship to build on from scholars both in English[8] and communication studies.[9] The primary distinction among these scholars is that some engage in rhetorical criticism of environmental issues (environmental rhetoric), while others mobilize ecology as a science and/or a metaphor for theorizing rhetoric and materiality (ecological rhetorics), which can but does not necessarily address environmental issues. In the introduction to their book on ecological rhetoric, Bridie McGreavy and her colleagues situate ecology as a trope that is "continuously reconfigured through repetition and difference, expressing rhetoric's creative potential."[10] In tracing the turns of ecology through rhetorical history, they identify three traditions of ecological rhetoric: (1) constitutive rhetoric, especially via articulation and transhumanism; (2) ecological models of composition and rhetorical i (3) and place-based methodologies in rhetorical fieldwork, or wha called engaged rhetorical practice. In their theoretical and metho training, many scholars of ecological rhetoric are associated with rhetoric of science, technology, and medicine (RSTM), a subfield of rhetoric that applies rhetorical analysis to scientific discourses and practices.[11] But while Kelly Happe has most recently studied racial formations and gene science, it is fair to say that intersectional views of science, technology, and environment are an area of growth for RSTM and ecological rhetorics at large.[12] As this project extends the rhetorical work at the junctures of science, environment, and coloniality, it argues two things: First, and with a few notable exceptions, ecological and environmental rhetorics via RSTM have largely neglected critical-cultural theories as a gauge for the effects of the sciences and politics it studies. In this regard, RSTM could productively follow the lead of environmental communication scholars like Phaedra Pezzullo, Kathleen de Onís, and Bridie McGreavy.[13] Second, to ensure rhetorical scholars conduct this work appropriately, the subfields of ecological rhetorics, RSTM, and de/colonial rhetorics should maintain critical conversations about their intersections. Encouraging ecological and RSTM scholars to read more broadly into postcolonial and decolonial science studies is a good start. A pressing question then becomes which theories and methods can ecological and RSTM scholars most productively build on to do this coalitional work, particularly at a time when traditional approaches to environmental rhetoric and new materialist-inflected ecological rhetorics have largely pursued other lines of inquiry?[14]

More than fifteen years ago, J. Blake Scott outlined a cultural studies approach to rhetoric of science (RoS) that has since become canonical among scholars of rhetoric of science, technology, and medicine (RSTM). By identifying the affordances and limitations of three waves of RoS—scientific rhetoric, scientific discourse and genres, and public scientific rhetoric—Scott argued that RSTM needed new hybrid approaches to account for science's broader conditions of possibility and to target opportunities for intervention by evaluating science according to its effects.[15] Scott called this a rhetorical-cultural approach because it studied science in equal relation to other cultural practices that mattered for the politics of each particular case. The characteristics of this approach thus fused an emphasis on the material-discursive to an openly critical stance against the negative effects of technoscience. In following a particular materialist vein in RSTM scholarship through Richard Doyle, Dilip Gaonkar, Bruno Latour, Donna Haraway, and Michel Foucault, Scott was one of the earliest to demonstrate that a more-than-human version of RSTM did not have to sacrifice politics in order to theorize materiality.[16]

Now more than fifteen years later, in the context of what some scholars have called a new materialist turn,[17] one has to wonder what RSTM would look like if while embracing this turn it would also have taken up the reading and writing habits of scholars like Blake Scott, Ellen Cushman, Sylvia Wynter, Donna Haraway, and Chela Sandoval. Like these scholars, can RSTM learn to think with Chicana/os and Latina/os, African Americans, and Indigenous peoples as much as we think with kinetics, epigenetics, and ecology? Can RSTM truly aspire to become the illegitimate children of the technosciences they study rather than merely their anthropologist? Or as Haraway and Stengers have put it most recently, in fusing the political with the material aspects of rhetoric, can RSTM scholarship make explicit commitments to be for some worlds and not for others?[18] Such efforts are not a letting go of commitments to scientific practices but a connectivity among commitments to ethical scientific practices as well as democratic ones. In RSTM's efforts to practice the nonmodern, surely it would be more durable and portable if it also practiced a version of noncoloniality.[19] Such efforts are not a rebirth of the cultural relativism of the science wars. These efforts are instead a renewed embrace of the notion that gender, sexuality, racializations, and more are key analytics for democratic politics at nearly every scale. Such efforts to co-labor with sciences and the cultural politics of democracy might go quite a way in helping RSTM answer the question of how rhetoricians can be pro-sciences without falling into the trap of scientism, because while sciences are RSTM's wheelhouse, as a rhetoric RSTM cannot think with sciences alone.

Some scholars may argue that embracing de/colonial analytics will fracture commitments to rhetoric, but closer examinations reveal these commitments are strong and perhaps more dynamic than ever. In Blake Scott's argument for hybridity, for example, he did not sacrifice the rhetorical for the cultural. In fact, he specified a mode for rhetoric—its unique ability to analyze the production and reception of texts, the ideological underpinnings of arguments, and its attention to the details of language—and a mode for the cultural—including its analytical emphasis on the political as an everyday cultural practice that values critique and tactical strategies of intervention.[20] Ultimately, in arguing for a hybrid approach, rhetorical-cultural critics do not abandon a radical sensitivity to context and situation in their grounded approaches to rhetorical analysis, yet at the same time the assessment of rhetoric should be based on its broader functions and effects, including material effects for embodied subjects. Importantly, the critics are part of this process and therefore *partial* in their material-discursive practices to attempt to create some worlds and not others.[21] As I argue throughout, moving from hybridity to pluriversality also does not fracture commitments to rhetoric but rather situates rhetorical analysis with a unique and viable hermeneutic and heuristic function. When practiced through notions of the pluriversal, rhetorical analysis helps take account of the agencies of materiality through colonial difference—that is, tracing ecologies of practice across heterogeneously entangled worlds under the power differential of coloniality.

Other rhetorical scholars may argue that commitments to decolonial praxis will fracture RSTM's commitments to study scientific practices, but a new generation of RSTM scholars has proven quite durable and portable in this Latourian manner of speaking.[22] To demonstrate, I point us to a number of collaborations among scientists and RSTM scholars, including publications in major scientific journals,[23] winning federal grants from the nation's major scientific institutions,[24] field collaborations with ecologists and medical practitioners,[25] and the burgeoning intersections of rhetoric and science communication,[26] all of which are informed by the intersections of rhetorical, cultural, new materialist, and increasingly de/colonial analytics. Given the depth and variety of this work, rhetorical scholars should not be pessimistic about the future of rhetoric-science collaborations at any level. In fact, one irony may be that in our post-truth political context, scientists are more willing than ever to listen to rhetoricians who can map, articulate, and intervene in the noise circulating in political worlds.

Ecologically minded RSTM scholars should also be encouraged by the engagements of rhetoricians and science and technology studies (STS) scholars who are beginning to seriously consider questions of epistemological and ontological plurality as essential for their work. To give just one example, in Scott Graham's exchange with Samantha Frost's work on biocultural creatures, we see a commitment to think with the sciences as one form of epistemology among many.[27] This project aligns with the sentiments there that RSTM should mobilize science as a form of inquiry *and* as an object of critique. As they note, the practices of science too easily become the practices of capitalist accumulation and population control; they too easily exert a mission creep to quantify the qualitative in ways that reify a rhetoric of supremacy.[28] What these conversations make clear is that if RSTM and STS more broadly think with the sciences alone, what's at stake is not epistemological authority but our ability to engage in practices that sustain common worlds—the very exigence to build trust across differences in our post-truth era. Like Graham, Frost, and others attest, a rhetorical-cultural model identifies and works against the tendency of researchers and policy makers to reduce environments and bodies, communities and cultures, to the molecular imperative so as to meet the demands of quantitative measurement, population control, and profitable marketing.[29] RSTM scholarship should be partial and should be against that world. Frost suggests that we explore how a deliberately multiscalar account of the environment might offer resources for political resistance and/or radical democracy.[30] I suggest that rhetorical-cultural analysis has given rhetoricians the building blocks of these multiscalar accounts, but what is also needed is an analytic that captures a plurality of rhetorical hermeneutics in ways that decentralize Western onto-epistemologies, and finally begins to consider world-building practices from elsewhere and otherwise.[31] One way for ecological rhetorics, and RSTM broadly, to do this is to embrace the pluriversal and embrace border thinking as another version of nonmodern praxis that works within and against the power differentials of coloniality. Ecological and STS rhetorical scholars need not make big claims to decoloniality in order to do critical border work; rather, through a nonmodern embrace of the pluriversal, scholars may engage in a critical analysis of the everyday colonial matrix of power through rhetoric.[32]

What would it mean for RSTM to embrace pluriversality in the sciences it studies, teaches, and engages? It certainly means moving from rhetorical approaches to science studies and engaging the key terms already in conversation with de/coloniality—nonmodernity, local histories, ecology, power/knowledge, institutionality, and intersectional analyses of race, gender, sexuality, and more. But as Wanzer-Serrano teaches us, the specific value of de/coloniality is a differential movement among various identity and class categories that yet find a common origin in the evolving legacies of colonial social orders. So while engagement with identity-based theories and critical-cultural categories generally is necessary and vital, tracing their roots to coloniality offers an intersectional analytic that allows for movement from "a politics based on identity" to "an identity based on politics."[33] It is exactly this form of differential movement to allow for intersectional analysis that RSTM could use to think through Euro-American notions of science and politics and affirm a world in which many worlds can fit, including a world of many sciences. Such movement would not only allow RSTM and ecologically minded rhetorical scholars to speak with decolonial science studies scholars more effectively, but it would also help do the necessary differential and coalitional work demanded by our moment.

To do this intersectional work, this book extends both the theory and practice of rhetorical-cultural analysis through a sustained engagement with feminist science studies and de/coloniality at a borderlands site of local climate politics. In doing so, this project recognizes that it is no coincidence that scholars across Latinx rhetorics and RSTM have come to similar conclusions about the function of rhetoric to mediate cultural relativism through rhetoric's radical contextualizations and democratic impulses. For example, sensing a similar disconnect between theory and practice in decolonial scholarship, Wanzer-Serrano notes that "decolonial theorists can lack a level of rhetorical specificity that, ironically, makes it harder to explore the geographic and embodied discourses they see as paramount."[34] Viewing

So much preamble w/o demonstration

rhetoric as a corrective to these trends, he submits, "De/colonial scholarship can benefit from a more rhetorical orientation that is highly attentive to practices of radical contextualization, sociohistorical contingency, and the situatedness of public discourses and activism."[35] Similarly, in their call for RSTM and feminist science studies to read more broadly, Booher and Jung note the long-standing relationship between scholars like Donna Haraway and Chela Sandoval.[36] And indeed, it's worth noting that Chela Sandoval has written about "New Sciences" as an alliance among her theory of oppositional consciousness and Haraway's situated knowledges.[37] Sandoval also writes about technoscience politics as an anticolonizing cyberspace that can overcome what she calls the apartheid of theoretical domains.[38] As Haraway teaches us, it matters what thoughts think thoughts. It matters with whom we think. We must stay with the trouble.[39] So, I end this section with a few pressing questions for ecological rhetorics and RSTM more broadly: When will RSTM embrace Wanzer-Serrano's "responsibility for difference in material-semiotic fields of meaning"?[40] When will RSTM call on Haraway and Sandoval as a nexus of affinity with the same technologies of resistance, the same "love" in the nonmodern world, so we too can write against dominance and recognize our own coalitional possibilities?[41] *Climate Politics on the Border* argues that ecological rhetorics and RSTM have a crucial and partial role to play in their contributions to the broader collective of rhetorical scholars. To make this contribution, the project argues for an engagement with technoscience all the way to coloniality and offers pluriversal rhetorics for local histories and futures of climate politics.

AN ECOLOGY OF PRACTICES FOR PLURIVERSAL RHETORICS

Over the last decade, ecological rhetorics have engaged various theories of new materialism in order to better attend to the multiple ontologies of the more-than-human, the persuasive force of materiality, and indeed, the field-based practices of ethnography and ecological sciences.[42] As I noted, de la Cadena and Blaser's work makes it clear that ecological theories have much to contribute to the pluriversal, but doing so successfully will require a much deeper investment into the intersections of ecological and decolonial rhetorics. My characterization of pluriversal rhetorics as the study and practice of rhetoric as an ecological (material and relational) system that functions across heterogeneously entangled worlds against the power differentials of coloniality is an attempt to do just this. In this way, pluriversal rhetorics are intended to be a critical intervention into what has been called a new materialist turn and more specifically an engaged rhetoric of science.[43] As Caroline Druschke recently argued, critical investments into ecological rhetorics and new materialism cannot be done solely through Heideggerian modes of being and practicing; they must largely de-link from Western assumptions and begin to critically listen to and engage with the praxes originating from the Americas.[44] Yet, the pluriversal rhetorics I argue for here cannot entirely separate themselves from Western modes of thinking and practicing. Rather, the pluriversal is an attempt to de-privilege, and thereby reconsider the relations of, Western theories and practices by engaging in border thinking and practice. So, for pluriversal rhetorics, it makes sense to think with these nonmodern philosophers as an internal critique of Western modernity albeit now with the added dimensions of decoloniality. In taking a cue from de la Cadena and Blaser, this section engages the relationship among ecological rhetorics and Isabelle Stengers's notion of an ecology of practices in order to understand how rhetorical practices compose, assemble, and hold together these ecologies as acts of creation that resist one-world extractivism and articulate pluriversal visions of the future.

Local climate politics is a productive site for this work because thinking with climate is a material-relational excess that cannot be fully contained by the productive forces of modernity/coloniality. As I noted in the introduction, the concept of climate is profoundly regional and rhetorical—it connects material conditions (environments) and public discourses (meaning-making) through the situated regional norms of creating worlds together (politics). In this sense, climate is always emplaced and experienced multiply. Climate is not a single entity but rather is itself better characterized as a pluriverse—a world among worlds. To make this argument more concrete, each chapter in part 2 begins with an evaluation of climate science according to its effects and then maps rhetorical practices around specific climate and weather events—particularly flooding events—as pluriversal happenings. Importantly, mapping rhetorical practices around extreme weather events is not done for the sake of cultural comparison but rather for an analysis of the constitutive practices that emerge in their specificity and with other practices also responding to these events.[45] For example, in responding to an extreme flood, many publics have a common interest in resilience, and yet what constitutes resilience is a site of divergence, an excess of interests, understandings, and practices that are not the same even as they emerge together. Shared interests in resilience, development, place-making, and adaptation practices emerge together but not without the excesses and uncommons marked by differential ways of knowing and doing within the colonial matrix of power. To engage the pluriverse as an analytic useful for conceiving of practices across heterogeneously entangled worlds, rhetorical science studies scholars will find it productive to think with one scholar who has had a wide influence among decolonial science studies—Isabelle Stengers.

Stengers is not as well read in rhetorical studies as some of her contemporaries in feminist science studies, but she offers a valuable materialist language that is particularly approachable for pluriversal rhetorical inquiry. This is especially true of her notion of an ecology of practices, which de la Cadena and Blaser center in their definition of pluriversal, but rhetorical scholars have only treated marginally.[46] Like many scholars in feminist science studies, Stengers is critical of simplistic views of science and scientific methodology decontextualized from their radically contingent practices of knowledge-making. Instead, Stengers forwards the notion of an ecology of practices, which traces how specific sciences interact with other equally valid knowledges and practices that matter because they shape the future.[47] Much like Haraway's notion of situated knowledges, the purpose of ecology of practices is not to abandon science, nor to see it in post-factual relativist terms characteristic of the science wars. For Stengers, each science differently mobilizes its own ecology of practices, its own craftwork, to create actual and particular claims on materiality and the way it functions. The problem with *Science* for thinkers like Stengers and Haraway is the way in which it maintains its pretentions to universality and full objectivity, which too often manifests as scientific imperialism and/or a colonialization of other modes of knowing.[48] To know a *science*, they argue, one has to know its specific and radically contextual practices, its situated knowledges.[49] The point of an ecology of practices is not to be simply tolerant of other ways of knowing, but to map practices that may coexist and co-labor together in a negotiation of interests.[50]

It helps to understand that Stengers's ecology of practices has its intellectual roots in Deleuzian rhizomatics, where nonhumans are causes for thinking, and thinking is "not accepting the state of affairs in the (majority) proposed terms [but] it always means the emergence of diverging minorities."[51] For Stengers, ecology captures the ways in which practices are irreducibly defined by relationships, by the association between the ethos of a practice and its *oikos*, "not only the matter-of-fact environment but the way it defines its relation with other practices and the opportunities of the environment."[52] Because an ethos can only be defined in relation to its *oikos*, changes in environments are changes in relations and events that transform practices. For Stengers, the term *practice* is not necessarily descriptive, but speculative. Practice implies situated belonging, and a radical immanence where particular practical settings cannot be abstracted to general relations that would "logically or consensually impose itself outside this setting."[53] As Haraway might say, there are no "God tricks" here. Rather, practice as a situated belonging refers to the constraints and obligations of our relationships as creative acts. Ecological relations constrain and oblige practitioners in ways that cannot be predicted, yet through an ecology of practices we can learn about these relationships and their possible creations. And as Stengers writes, "Learning here is always local because the rhizomatic connections practitioners are able or unable to forge do not obey general rules and reasons."[54] In terms of the sciences associated with climate and weather, one crucial question becomes, how can these scientific communities address different allies than their traditional ones—the state and industry—who continually betray this alliance? As Stengers claims, to do so they have to learn how to present themselves in nonmajority terms.[55] To do this, Stengers describes how an ecology of practices always carries the possibility of the creation of a specific cosmos—not a single universal ideal order, but rather that which is "constituted by multiple, divergent worlds and to the articulations of which they can be capable."[56] These possible articulations as partial connections between practices are cosmic events, "a mutation which does not depend on humans only, but on humans as belonging, which means they are obliged and exposed by their obligations."[57]

To help concretize these ideas, Stengers gives the examples of a virus and a flood. In an ecology of practices, she writes, "it is not an objective definition of a virus or of a flood that we need, a detached definition everybody should accept, but the active participation of all those whose practice is engaged in multiple modes with the virus or with the river."[58] Thus, all those who participate with the flood, who practice with the flood, matter for political thinking. For cosmopolitics, she writes, the question is twofold: "How to design the political scene in a way that actively protects it from the fiction that 'humans of good will will decide in the name of the general interest?' [And] How to turn the virus or the river into a cause for thinking?" For Stengers, political thinking has to proceed "in the presence of" those who would be affected by the decisions, and they cannot a priori be thought of as disqualified and unable to contribute to a common account.[59] For Stengers, cosmos is an operator of "putting into equality" in opposition of any notion of equivalence: "equality does not mean that they all have the same say in the matter but that they all have to be present in the mode that makes the decision as difficult as possible, that precludes any shortcut or simplification, any differentiation a priori between that which counts and that which does not."[60] For a flood to become a cause for thinking, the onto-epistemological realities of the more-than-human cultivate an immanent attention "to presences that are or can be but do not meet the requirements of modern knowledge and therefore cannot be proven in its terms."[61]

It is through Stengers's conception of an ecology of practice that rhetoric-ecology scholars can approach pluriversal rhetorics as a praxis of engagement with and for a world of many worlds. As Stengers writes, ecology does not equate to harmony or peaceful coexistence, but rather a web of interdependent partial connections that are co-constitutive even without a transcendent common interest, or even mutual understandings.[62] As a science and as a practice, ecology is concerned with the relations among living things and their physical environments, and across trophic levels, these relationships are characterized by symbiotic relations of all kinds (mutualism, commensalism, predation, parasitism, and competition). The affordances of an ecology of practices are therefore not just to multiply ontologies but to analyze and invent those symbiotic relationships.[63] As such, an ecology of practices is directly aligned with rhetorical-ecological scholars calling for trophic futures that attend to the scalar dynamics of matter and energy, the ways language manifests with materiality, and the relational complexity that characterizes the arts of living well together.[64] As the Haraway epigraph that began this chapter notes, protecting and caring is what entitles practitioners to participate in a process of composition as a site of knowledge-producing and world-building practices. As Stengers has written recently, the primary question then becomes one of relevance, or how and under what conditions are relations among practitioners relevant to the situation at hand? What is relevant is not solely based on reason or persuasion, but rather what is relevant for staying with the trouble across contact zones and among what is typically defined as mutually exclusive.[65] De la Cadena and Blaser's centering of an ecology of practices in the pluriversal provides a starting point for characterizing pluriversal rhetorics as ecological (material and relational) and heterogeneously entangled with many onto-epistemological worlds that matter for decision-making.

Pluriversality can help rhetorical science studies scholars address a profound silence over parallels among modernity and coloniality in their studies of science, its rhetorics, and its politics. In providing an analytical and inventional approach capable of conceiving an ecology of practices under the power differentials of coloniality, the pluriversal is capable of responding to the disruptions of climate breakdown as a reconsideration of the relations of divergent yet entangled worlds.[66] Latour famously argued for the concept of the nonmodern, but in de-linking science with modernity, his work has done little to de-link science from colonialism or systemic racism, which would seem to further the project of the nonmodern.[67] It is in following the lead of feminist science studies scholars like Haraway, Harding, and Stengers and decolonial scholars like Escobar, de la Cadena, and Mignolo that one can witness the broader potential for the pluriversal. This is partly why Mignolo continually asserts the value of Stengers's de-linking of cosmopolitics from Kant, and Latour's position of nonmodernity as akin to decolonial cosmopolitan ideals.[68] As Mignolo writes, "One could even think of

Latour as a decolonial European thinker and, as such, see his horizon as the imperial rather than the colonial side of coloniality."[69] Indeed, in his articulation of a cosmopolitan localism or decolonial localism, Mignolo asserts that "Latour's pluriversal cosmopolis and decolonial cosmopolitan localism could join forces in promoting the coexistence of cosmopolitanism, with modifiers and in parenthesis."[70] Pluriversal rhetorics are my version of this coexistence, and the local climate politics of San Antonio are my site of pluriversal rhetorical praxis. To begin to address the rhetorical tools needed for pluriversal praxis, and indeed to think with climate science and climate change as enactments of decolonial border thinking, I first need to make explicit connections among place, publics, and politics for pluriversal rhetorics.

POLITICS: PUBLICS, PLACE, AND POWER IN PLURIVERSAL RHETORICAL PROJECTS

In asserting that matters of climate cannot be reduced to science and must include the knowledges and practices of situated publics, this project engages in a place-based and publics approach to rhetoric.[71] Public rhetorics are particularly appropriate for emplaced, ecological, and engaged projects because "being there" in the field allows scholars to capture the everyday, vernacular discourses in addition to the human/more-than-human ones so important to interdisciplinary fields like RSTM. By inhabiting the sites of their study, emplaced and engaged projects offer access to a wider range of texts, practices, and experiences than one might trace otherwise.[72] As Candice Rai notes, such an approach helps capture the places of persuasion, or how rhetorical invention "emerges from and is brought to life by particular environs already in the world," and these environs and situations are characterized by multiple and interrelated meanings of "place."[73] Yet as many of these scholars note, it is difficult to study power and politics through field-based rhetorical methods alone since power, by its nature, operates through exclusion, and gaining access to sites of significant power always remains a challenge. Thus, field-based projects should not sacrifice theoretically sophisticated standpoints for the saturations of another interview, observation, or survey. Rather, as much as engagement in fieldwork means "being there," researchers can still benefit from distance, too, in order to be appropriately critical of the power dynamics of our sites of study.[74]

As an emplaced, engaged, and ecological project with a circular and living view of history, this book speaks to the ways rhetorical histories can function as a form of engagement. Furthermore, as a rhetorical project in the borderlands of San Antonio, Texas, it also uses location to ask critical questions about de/colonial inventions in relation to climate and weather events: How have extreme weather events in the past mobilized publics to consider relations among divergent worlds within the differential power of coloniality? How have these relations among worlds been performed through public discourse and which ones were persuasive through local democratic structures of governance? What are the functions of technoscience in composing these relations and influencing political decisions? What follows in this section is a methodological discussion of how rhetorical techniques of *topologies* and *tropologies*—or commonplaces and their un/expected turns—can be understood as both rhetorical heuristics and hermeneutics for emplaced, ecological, and pluriversal projects. First, I discuss *topologies* as public (shared), placed (material), and ecological (relational) rhetorical techniques that function in multiple dimensions. Then, I discuss *tropologies* as un/expected turns of embodied language use. Next, through an analysis of decolonial pluritopic hermeneutics I demonstrate the value of these techniques for decolonial and pluriversal projects. Throughout, I argue for a treatment of *topologies* and *tropologies* as microanalytics and dynamic, nonlinear, and transductive ways to trace discourse and practice as mediating functions of place that captures a plurality of relations among worlds.

How diverse and competitive public bodies engage in a measure of collective action through deliberation has always been a fundamental concern in rhetorical theory and practice. While rational talk and debate has historically been the mechanism by which the legitimacy of public decisions is made manifest, scholars of public rhetoric have increasingly turned toward studying everyday and vernacular discourses as a way to correct for some of the limitations of overidealized public spheres. Broadly, this research suggests that everyday publics, vernacular discourses, and democratic subjectivities are radically situated, responsive to contextual dynamics, and emergent through discursive circulation and practice.[75] Some of this work has itself been critiqued for seeking "ideal democratic subjectivities" and as a corrective, rhetorical scholars have argued for an investigation of democratic subjects as they emerge through everyday rhetoric and materiality in situ. As Candice Rai explains in her ethnographic study of a Chicago neighborhood, in a publics approach to rhetoric and materiality, one examines "the co-constitutive relationship between (democratic) rhetoric and materiality; how rhetoric latches onto ideologies in the contexts of its use; the rhetoricity and agency of materiality itself . . . through which people are transformed into rhetorical subjects that emerge 'within a specific apparatus of production.'"[76] While the influence of feminist STS and decolonial scholars are my primary means for theorizing materiality in climate politics, this approach to public rhetoric and materiality as co-constitutive in everyday democratic practice provides a unique and textured manner to trace the emergence of placed democratic subjectivities responding to and in anticipation of extreme weather events.

My investments in the public work of rhetoric are based on a few key rhetorical tools that I understand as microanalytics capable of interpreting and intervening in the publics and politics of San Antonio—namely topoi and tropes.[77] Typically, an analysis of public rhetoric would study topoi as commonplaces or common topics of public discourse, perhaps alongside notions of atopos as the loss of what is commonplace or shared. Rooted in Sophistic traditions, topoi have enjoyed a long history of reinvention in rhetorical studies.[78] Classically, topoi could refer to common discursive "places" like lines of argument, lists of questions, or argumentative forms; alternatively, they could refer to publicly shared knowledge specific to a time, a place, and a community. For contemporary rhetoricians interested in materiality, it is the latter definition of topoi as situated, emergent, and in circulation that provides rhetoric with an attention and responsiveness to everyday life. Working from Ralph Cintrón and Carolyn Miller's situational definitions of topoi as "storehouses of social energy," and "a space, or a located perspective from which one searches," rhetorical scholars John Muckelbauer and Candice Rai argue that topoi provoke a kind of inhabitation wherein one would need to be familiar with a particular place in order to properly understand or wield rhetorical arts.[79] Most recently, Casey Boyle has shown how the critique that topoi are plagued by fixed repetition and also unwieldly multiplicity is actually an affordance when understood as situated, temporal, and mediated activity with novelty across repetitions. For Boyle, understanding topoi as always potentially novel leads him to argue for commonplaces as an "immanent practice of selecting and collecting texts with a goal of transducing place."[80]

Indeed, the notion of topoi as provoking an inhabitation and immanent practice that transduces place in ways both usual and novel, familiar and unfamiliar, is partly how Lynda Walsh and Boyle distinguish topoi from topologies. They write, "Topology combines the classical rhetorical strategy of topos with nonlinear logic to yield a new model of discourse that is (a) transductive in its workings across concrete examples to illuminate structure, (b) responsive to the contingencies of the discourse situation, (c) generative of power dynamics that help shape that situation."[81] Unlike topoi, topologies remain responsive to situational contingencies and power dynamics while still accounting for structure across situations. Topologies are un/familiar. They highlight the co-constitutive relationship among topoi as the common, and kairos as the opportune spatial/temporal moment for rhetorical performance.[82] It is the situation-specific relationship among topoi and kairos that rides topologies with novelty because here, and against Aristotle, topologies are understood as nonlinear, nonlogical, and yet still a way to trace and invent discursive practices.[83] Thus, the primary features that distinguish *topoi* from topologies are that topologies track the beliefs, values, and norms of publics as a process of immanent versioning and multidimensional folding rather than the linear synthesis common to logic and dialectics. In other words, topologies capture the rupture of logics to politics (and of common to universal) as a multiplicity of situational folding and articulating. There is no naturalizing or essentializing here, only what we might call an ecology of rhetorical practices that are traceable as inventional performances for the emergence and legitimization of situated decision-making. In this project, the topologies of shared sacrifice, divergence and diplomacy, and inhabitation as place-keeping function as un/commonplaces of response to extreme weather events—a rhetorical versioning of climatic response-ability. These un/commonplaces function transductively as both historical hermeneutics and contemporary heuristics for unique and contingent situations.[84]

As the Haraway epigraph to this chapter notes, world-building practices and joint knowledge production build on topologies, but also take unexpected turns, or what rhetoric refers to as tropes, or the reoccurring figurative use of language that means other than what it might ordinarily signify. Like *topoi*, tropes can be common or novel depending on situational dynamics that constantly turn or alter meanings through inventional practices that question what situations can mean here and now. Rhetorical work on tropes (metaphors, metonymy, hyperbole, and more) has argued they are ubiquitous patterns and forms that can be recognized across domains as doing persuasive work by epitomizing arguments[85] and offering inventional resources for seemingly endless novelty.[86] New materialist rhetoricians have also drawn parallels among the etymology of trope as a "turn" to argue they are affective modes of ecological relationality and change, and they index this understanding through various terms such as "heliotrope,"[87] "tropic," and "trophic."[88] Thomas Rickert and Diane Keeling and Jennifer Prairie have suggested that tropological "turning" is useful for engaging rhetorical addressivity and responsivity in the way nonhumans interact with one another and with their environments.[89] Thus, the unexpected turns of tropologies also help account for novelty within repetition as modes of rhetorical and ecological encounters and relationality. Topologies and tropologies are thus relational world-building and knowledge-producing practices that are particularly appropriate for borderlands spaces. As Gloria Anzaldúa teaches us, borders are both figural (tropological) spaces—"una herida abierta [an open wound] where the Third World grates against the first and bleeds,"—and literal (topological) places "wherever two or more culture[s] edge each

other, where people of different races occupy the same territory, where under, lower, middle and upper classes touch, where the space between two individuals shrinks with intimacy."[90] Like borders, topologies and tropologies both perform and track those spatial, relational, and ecological modes of inhabitation with emergent and un/expected lines of affinity across differences worked out in practice.[91]

Recently, *topoi* have come under critical scrutiny again from new materialist and social constructivist camps. The former critiques the logical foundations of *topoi* as a neglect of the affective and material dimensions of rhetorical practice; in this vein scholars often offer up the concept of *chora* (Greek: kora) as an alternative, though more work needs to be done on *chora* in regard to politics.[92] From the social constructivist side, *topoi* have been critiqued as moored to location, constitutive of static identities, and ordered for the purposes of flexible accumulation. As an alternative, social constructivism offers what Michel Foucault calls *atopics*—the "loss of what is 'common' to place and name."[93] Yet these assumptions of *topoi* as merely local, logical, and static are misguided, and they certainly do not retain the necessary relationship to *kairos*. Rather, as Joshua Reeves has written, the *atopos* as the out of place, strange or uncommon, is perhaps less about placelessness than it is about "the peculiar position from which an extraordinary rhetorical act compels its audiences to reencounter the world."[94] Reeves's notion of *atopos* as a reencountering through rhetorical acts might also be described from a topological understanding as a fold or divergence in the creation of a topology. The atopic fold, as it intersects with other topologies, is the divergent yet productive space of encounter. As I explore more specifically in chapter 3, *atopos* is a site of divergence in articulation, or the becoming with what one is not, while also not becoming what one is not.[95] One value of topologies for public rhetorics then is their ability to capture both *topos* and *atopos* in the invention and analysis of multiple community commonplaces with their folds and divergences, particularly under shared *kairoi* like a city experiencing an extreme weather event.

The *topos/atopos* binary can be productively ruptured through topologies wherein both common and uncommon places unfold in multiple dimensions and specific moments where partial connection and divergence is always already taking place. In her study of Ecuadorian national identity, Christa Olson discusses the territorial implications of *topoi* and asserts they are "places of return in changing circumstances" or "placed confluences."[96]  Return under changing circumstances elegantly articulates the un/familiar capacities of *topoi*, yet from the view of topologies, confluence connotes a kind of synthesis or union that displaces their multidimensional folding as a process of invention. In contrast, rhetorical topologies are less characterized by confluence than they are by networks and whose *topoi* function as terminals with elastic ties to other terminals in the network. As Nathan Stormer and Bridie McGreavy note, from an ontological view of rhetoric *topoi* as commonplaces may be insufficient to be a methodology by themselves; instead, *topoi* are more like Burke's notion of termini, which produce different encounters with phenomena.[97] Building on this work is Byron Hawk's notion of sphere publics wherein spheres are not just a set of static nodes but the emergent and co-productive properties of networks that help describe local, fragile, and complex formations that network theories cannot account for.[98] Rhetorical topologies might understand these emergent and co-productive sphere publics not as convergences but as linked termini or sites of partial connection that are also sites of divergence in a broader ecology of discourses and practices.

Let me provide an example. San Pedro Park in San Antonio is one of the oldest city parks in the nation. It could be understood as a *topos*, a public place with widely shared knowledge of its territorial boundaries, its temporalities (that is, when the spring-fed pool is open in summer; when the March for Science happens), and its own sense of community. But a topological approach understands that public rhetorics rarely have a discursive confluence in the way a physical park might suggest. Rather, the park is a site of endless mediation and multiplicity through its encounters with various publics. As such, the park is less a place of confluence than it is a site of partial connection that is also characterized by excess and divergent understandings of what the park can mean. For many, the park is a historical landmark; for others, a critical public space for protests and recreation; still for others, a site best made available for private development, and so on. To put it in a Latourian manner of speaking, shared matters of concern rarely contain shared understandings of those matters. Yet, the practices of protecting shared concerns, of caring for them, is not immobilized by diverse understandings. It is just the opposite. Diverse understandings of San Pedro Park can be traced through topologies and tropologies as discourses and practices that fold, stretch, turn, and provide shape to various understandings of shared spaces. While these discourses and practices are often characterized by parataxis and/or divergence, they are also always possibly articulated together (a sphere public) for the purpose of political action—stopping the destruction of the park, for example. Un/commonplaces and their un/expected turns help account for a plurality of commitments, knowledges, and engagements from which collectives emerge for world-building, care-based practices. It is this capacity to trace public discourses and practices with their tactical overlaps and divergent excesses that make topologies and tropologies such remarkable tools for the contingencies of coalition building.

De/colonial theory has also engaged with similar nonrational and nonlinear pluriversal techniques, but it stands to benefit immensely from rhetorical analysis via topologies and tropologies. One might begin with Madina Tlostanova and Walter Mignolo's development of a pluritopic hermeneutics, which builds on Raimon Panikkar's notion of diatopic hermeneutics as "an art of understanding by means of crossing spaces or traditions (dia-topoi), which do not have common models of understanding and understandability."[99] Pluritopic hermeneutics emphasizes connecting one's own body and experience to the social, political, and ontological dimensions of theorizing while understanding that "other truths also exist and have the right to exist, but their visibility is reduced by the continuing power asymmetry, which is based on the coloniality of knowledge, power, being, and gender."[100] As a theory of power, coloniality operates on quite a different level than rhetorical topologies and tropologies, but as pluritopic hermeneutics suggests, that does not mean they are oppositional. They are just other truths that also exist. But the affordance of their interactions together would seem to help the microanalytics of rhetorical analysis come to grips with power via the colonial wound, and help the macroanalytics of de/coloniality get grounded in situation-specific contexts that work not just hermeneutically but also heuristically as inventional practice for emerging contexts. Indeed, ecological and decolonial practitioners are beginning to argue that these kinds of interactions are essential for transhemispheric de/colonial politics. For example, Arturo Escobar notes that Northern transition discourses need further interaction with the decolonial and postdevelopmental paradigms of the Global South, and the Global South needs to further engage with technoscience, and especially ecological designs: "By connecting to each other," he argues, "they might extend like rhizomes, possibly emerging into *local and regional topologies of partially connected worlds*."[101]

The negotiation of ecological and de/colonial public rhetorics among diverse cultures and species with long histories of colonial contact is what San Antonio has to offer for rhetorical studies. At this site, by comparison, *chora* as "a generative and orchestrating background" requires more specificity.[102] When treated in isolation, notions such as *atopics* induce a double inflection through an a/topic binary that does little to help rhetoricians come to grips with politics. Pluriversal topologies and tropologies demand that rhetorically inflected analyses be situated through radical contextualization, be attentive to material-discursive performances, and be political through the world-building practices of articulation. The promise of tracing topologies and tropologies through local climate politics then is to both analyze and invent the discourses and practices where a broader sense of ethical and just outcomes may be possible in a world where climatic and political stabi can no longer be assumed. Commitment and engagement in an emerging collective open to caring and crafting noninnocent and contestable locations of research full of human and unhuman actors building on un/common places and taking un/expected turns—this is a positioning also codified in the last phrase in the book's title: On the border.

ON THE BORDER: BORDER THINKING FOR PLURIVERSAL RHETORICAL PRAXIS

To live in a minority-majority city like San Antonio is to co-exist and co-labor within multicultural contact under coloniality every single day—to culturally, linguistically, figuratively, and physically find oneself in practice with multiple worlds. To understand the political work of climate in this city as an Anglo-Celtic, western US–based, settler scholar, who lives with white privilege, and who is not native to San Antonio, is to embrace the status of an outsider. For outsiders who prefer not to study from the territorial gaze of the disciplines,[103] but rather inhabit their sites of study, the appropriate methodological stance may be to literally and figuratively stand, practice, and deeply inhabit border spaces as sites of ongoing colonial wounds. For the purposes of this project, then, *on the border* does not signify an outside observation and description of borders using the tools of Western rhetoric but rather an inhabitation of border spaces wherein an outsider listens, learns, and engages in continuous dialogue with many human/more-than-human others to reconsider relations among worlds. Importantly, *on the border* does not connote a comparative method because too much of comparative methodologies ensure that the scholar/observer remains "objective" while Western knowledges remain in control. Instead, I view *on the border* as what Raimon Panikkar and Walter Mignolo call an imperative method: "the effort at learning from the other and the attitude of allowing our own convictions to be fecundates by insight of the other."[104] Rhetorical techniques enact an imperative and pluriversal method because rather than comparison, topologies and tropologies trace situated rhetorical practice through partial sites of connection under shared *kairoi* that are also sites of divergence. My own location as a rhetorical scholar in this book then is to embrace the risky position of practicing ecological rhetoric on the border so that I myself, like Donna Haraway and Chela Sandoval, may attempt to co-create lines of affinity with the same technologies of resistance, the same "love" in the nonmodern world— so I too can write against dominance and recognize coalitional possibilities.[105] *On the border*, then, connotes a methodological stance of pluriversal rhetorical praxis that is emplaced, engaged, and partial yet still connected through the power differentials of coloniality. What follows in this section is a discussion of how rhetorical border praxis might be conducted ecologically and pluriversally as a versioning of climate justice rhetorics. As the epigraph from Reiter suggests, throughout I argue that climate politics are best understood as entangled through and by coloniality, and pluriversal rhetorics are one way of dwelling in its interstices.

Few concepts have been as durable across rhetorical theory and practice as the various analytics that constitute border studies and border thinking. Much of this scholarship is informed by Gloria Anzaldúa's theorizing in *Borderlands/La Frontera, This Bridge Called My Back, This Bridge We Call Home*, and the recently released *Luz en lo Oscuro*.[106] *This Bridge We Call Home* and *Luz en lo Oscuro* were the last publishable writings from Anzaldúa's life, and they can be seen as a culminating gesture of her theorizing that is both specific to the Chicanx experience, and specifically for Chicanx politics, while also inclusive of any subjectivity that seeks to rethink existing categories and invent new political identities. Such is the brilliance of reading Anzaldúa—her readers enact her theory of nepantlera as an in-between and bordered space that refuses overly static identity categories, and continually invents new possibilities. Rhetorical studies have long been influenced by borderlands thought, especially Chicana feminism and its rhetorics,[107] Mexican and Mexican American cultural rhetorics,[108] border rhetorics, generally,[109] and Latin American rhetorics, broadly,[110] just to name a few. Most recently, Josue David Cisneros's book *The Border Crossed Us* takes on critical border thinking to extend the scholarship on border rhetorics. For Cisneros, border rhetorics are the expression of border thinking wherein borders are "both physical places at which nations meet and also cultural spaces at which national identity and community are demarcated."[111] As both material and semiotic, border rhetorics means thinking in multiple dimensions of discourse and practice of both dominant groups often invested in maintaining a status quo and in minoritarian groups often contesting social norms. Yet given the indeterminacies at the heart of identity and location, there is nothing essential or natural about borders; they are thoroughly rhetorical and therefore always contestable. Borders always function multiply and take shape through repetition and performance. This is why the promise of rhetorical approaches to borders and bordering as performances for in-between spaces always carry the possibilities of new identities and new political futures.[112]

Mignolo notes that border thinking is modeled on the Chicanx experience and can also be traced to the idea of African gnosis or gnoseology. Gnosis is knowledge in general—including *doxa* (opinion) and *episteme* (knowledge)—and border gnosis is this knowledge from a subaltern perspective.[113] Border thinking arises out of border gnosis in order to rupture the separations created among subject/object, knower/known, and more broadly, hermeneutics/epistemology. As Mignolo explains, "The goal of gnoseology is to erase the distinction between knower and the known, between a 'hybrid' object (the borderland as the known) and a 'pure' disciplinary or interdisciplinary subject (the knower), uncontaminated by the border matters he or she describes."[114] Such is the importance of inhabitation in pluriversal rhetorics—if I as an Anglo-Celtic American male living with white privilege am to engage border thinking as the space in which this logic could be thought through, then I must embrace and be embraced by the border matters I study. And this is true: Stemming from my experience with living and raising children in the southwestern borderlands for a decade and specifically in the heart of San Antonio for the last six years, border thinking is an everyday embrace. *Climate Politics on the Border* is an outcome of my own limited, partial, non-innocent, situated knowledges of climate politics in the Southwest and especially, San Antonio. It is impossible for those like me living with white privilege in the United States to think from the experience of the subaltern, but the analytic of coloniality, of inhabiting the colonial wound, is a coalitional standpoint (de/coloniality) where much depends on who comes together to shape the knowledges and practices of any given project.

As a city with multiple intersecting histories of colonial contact, San Antonio embodies border thinking as an art arising from the point of colonial contact and entailing the disruption of dichotomies by bringing subalternized discourses and practices into articulation with dominant ones.[115] For José David Saldívar, border thinking is a "geopolitically located thinking from the borderlands of Americanity and against the new imperialism of the USA."[116] With colonial power as its focal point of critique, border thinking and its extensions into border rhetorics are "positioned at the fissure of fault lines of Western hegemony and move between colonial and subaltern knowledges and voices, articulating a position that challenges both traditions and tensions."[117] While both postcolonial and de/colonial scholars often operate at former colonial sites, Darrel Wanzer-Serrano and Romeo García and Damián Baca usefully note two major distinctions among these groups: first, decolonial scholars critique postcolonialism for its centering of European histories, and for its emphasis on "post" when coloniality is still the most widespread form of domination and governance today.[118] Instead, decolonial scholars argue for an engagement in "epistemic disobedience" and "de-linking" from modernity/coloniality. Second, de/colonial scholars broaden the analytic focus to include all of modernity, and therefore provide an alternative to modernism, poststructuralism, and postmodernism.[119] The overarching movement of border thinking and decolonial theory in general is to "think in exteriority" with subalternized voices at the spaces of colonial contact in order to challenge power-laden formations of race, class, gender, sexuality, nationality, and more.

Importantly, however, both Mignolo and Wanzer-Serrano continually emphasize that when thinking in exteriority, there is still contact with modernity/coloniality, so the question largely becomes one of epistemic privilege rather than mere existence.[120] Decolonization is only one option among other possible futures, including de-Westernization, re-Westernization, reorientations of the left, spiritual options.[121] As Mignolo's reading of Stephen Toulmin's work on the European cosmopolis makes clear, a key question for the various ends of border thinking is the question of how scientific and technical practices mediate conceptions of the cosmic polity. Because scientific and technical practices were and still are complicit with the formation of colonial and imperial power and Euro-American expansion across the globe, a key question becomes, Can there be anticolonial, antimodern scientific discourses and practices that foster conviviality on a planet where many worlds can fit?[122]

These are the kinds of questions that projects for decolonial rhetorical science studies, or pluriversal rhetorical studies, might address, and they can start by contributing to the work already being done in postcolonial and decolonial science studies. The decolonial scholarship argues that the Anthropocene may very well offer opportunities for pluriversal worldings,[123] sciences from below,[124] and transition discourses with plural concepts of science and technology.[125] From this standpoint what is at stake with the climate crisis is not just a political economy, cultural practices, global climate policies, or geopolitical power struggles but "a crisis of a particular model of civilization . . . that of patriarchal Western capitalist modernity."[126] In other words, the climate crisis can become a cause for thinking about transitions away from colonial social orders. In this line of thinking, decarbonization, depatriarchicalization, and decolonization can partially be practiced through relocalization projects that decentralize food, energy, and economies to foster diversity of all kinds in place-making development projects. Though I am skeptical about uncritical calls to return to the local, or indeed calls to de-link fully from modernity/coloniality, de-centering Western conceptions of science and technology, and putting them into pluriversal contact with other situated knowledges is a productive area of conversation where rhetorical scholars across RSTM, de/coloniality, ecology, and border rhetorics can make valuable contributions. For example, in this book about San Antonio's climate politics, Escobar's notion of transition discourses—degrowth, the deep commons, postdevelopment, Buen Vivir, the rights of nature, and civilizational transitions—arise in the context of San Antonio's contemporary development rhetorics (see especially chapter 4).[127] Here it is important to emphasize that pluriversal projects are not based on models of cultural essentialism and authenticity that create overly static versions of place and identity. Rather, these projects foster a world of radical interdependence, immanent co-constitutiveness, that pluralizes the real and the technoscientific, even as they remain entangled through the colonial matrix of power.[128] This does not make scientific knowledge any less real; it only casts it in minoritarian terms that allow publics to address deficits of shared practice rather than deficits of shared knowledges.

Wanzer-Serrano and Escobar argue that decolonial transition discourses have significant alignment with civil rights movements and transnational networks that support climate justice on a global scale.[129] Thus, though there is a clear bias toward the local and place-based within de/colonial projects, the pluriversal can productively conceptualize these transnational support networks for local social movements, as many witnessed during the Standing Rock protests of the Dakota Access Pipeline. Similarly, Sandra Harding has argued that although environmental and climate justice movements have a history that has not emphasized de/colonial analysis, de/colonial thought has significant alignments with the collection of methodological strategies that flourished in the social justice movements of the 1960s. These strategies directly intended to create "'sciences from below' that articulated the conditions, needs, and desires of economically, politically, and socially vulnerable groups."[130] Despite the divergent genealogies and vocabularies among de/coloniality and environmental justice, both arise from the experience of the colonized, and therefore they represent another point of colonial contact that brings subalternized politics into articulation with dominant ones. So, while de/coloniality and environmental justice should not be conflated, one can clearly see how the siting of toxic waste dumps in primarily Black neighborhoods is an outcome that evolved out of the social orders of colonialism and slavery that still exist today. And, indeed, the case of Standing Rock suggests that one future of environmental and climate justice will continue to involve significant participation with Indigenous sovereignty and decolonial theory and practice.

The next two chapters of this book offer rhetorical, ecological, and de/colonial readings of extreme weather events in San Antonio's history, and in doing so, they identify a hidden history of San Antonio's own antecedent climate justice movements. As a political and social movement, environmental justice arose out of Warren County, North Carolina, in opposition to the systemic siting of toxic waste dumps in predominantly poor, Black, and rural areas of the South and across the country.[131] While the movement achieved some significant political victories in the early 1990s,[132] perhaps most significantly, it redefined environment as "the places we live, work, play, and worship"[133] as it also highlighted the importance of environmental racism: "racial discrimination in environmental policymaking, the enforcement of regulations and laws, the . . . targeting of communities of color for toxic waste facilities, . . . and the history of excluding people of color from leadership of the environmental movement."[134] Indeed, it is no mistake why many decolonial and environmental justice movements are beginning to coalesce around notions like a just transition to decarbonized ways of living well together—a point I elaborate on in the final chapter.

Hurricane Katrina in 2005 was an important transitional moment toward climate justice, but the extension of environmental justice into climate justice is most appropriately situated as arising out of the first Climate Justice Summit in 2001, and a report from the Black Congressional Caucus on African Americans and Climate Change in 2004. These groups defined key principles of climate justice such as CO_2 emission mitigation, protecting vulnerable communities, just transitions to renewable energy, procedural justice through community participation, and acting in the face of uncertainty.[135] Whether at the academic, elite organizational, or grassroots levels, one constant in climate justice work has been a focus on the unequal burdens placed on those who were the least responsible for climate change pollution. At the global level, this is often discussed in terms of international aid, but at the local level there is a clear focus on building capacity for more broadly shared just developments out of poverty, precarity, and injustice through the work of decarbonized climate adaptation. Indeed, it is the emphasis on climate adaptation as building capacity for reckoning with inequality that situates this book within climate justice rhetorics. De/colonial analysis would point out that the issues that climate and environmental justice movements seek to resolve have existed long before the 1970s and have evolved from colonial relations that still perpetuate systematic domination of people based on race, gender, sexuality, and more. If, as Naomi Klein says, climate change means everything must change, historicizing climate justice issues through the lens of border thinking, de/colonial analysis, and the pluriversal may help scholars gain clarity on the root cause of many systemic issues arising from colonial relations of domination. I believe these lenses will be necessary in a world where the colonial, extractive, and modernizing practices of the past will haunt all of us as we attempt to live through climate breakdown.

PLURIVERSAL RHETORICS UNDER THE NEW SOVEREIGN OF CLIMATE

Given our unprecedented situation of nine billion people on one destabilized planet, how can ecological rhetorics, climate justice, and de/colonial analysis even hope to create a modicum of equity in a world that is verging on a near future of ecosystem and civilizational collapse? Or said differently, and drawing from the work of Ralph Cintrón, what are the potentials of pluriversal rhetorical praxis when climate is the new sovereign and when it is easier to imagine a dying planet than a dying modernity/coloniality?[136] How exactly should critical rhetorical science studies scholars and practitioners proceed? As Sandra Harding asks, how should we attempt to foster many sciences on this planet?[137] These are the kinds of questions that haunt this project because the answers are also a plurality. Finding the right words matters. Archival work and fieldwork matter. Theorizing matters. Of course, activism and politics matter. Overidealizing Indigenous, Chicanx, African, and other traditions as free of domination, exploitation, and oppression won't help, but ignoring the global system of coloniality/modernity on which extractivism is based won't help either. If it is the relations among worlds that must be reconsidered through rhetoric, and if one cannot hope to fully heal the colonial wound, perhaps one at least can work to stop the bleeding for a while.

In order to punctuate the theoretical investments of pluriversal rhetorics, I end here with an attempt to identify some of its assumptions, capacities, key moves, and contributions. This chapter has detailed the intellectual genealogy of pluriversal rhetorics in relation to ecological and de/colonial rhetorics, which, I have argued, are an increasingly necessary relationship given the disruptions of climate breakdown. Both ecological and decolonial orientations to rhetorical studies contain certain assumptions that should be made clear:

- Modernity/coloniality are constitutive of our contemporary world.
- Local knowledges, histories, and landscapes are a primary means of questioning and resisting the extractivism of modernity/coloniality.
- Entanglement with the more-than-human is co-constitutive and interdependent with the human and cannot be thought as separate.
- Fostering worlds among worlds can co-labor toward a creative and more equitable co-existence.

Pluriversal rhetorics in relation to ecological and decolonial rhetorics also carry certain aims/goals/purposes:

- To foster worlds among worlds that work together to resist rather than reify one-world development models based on the power differentials of coloniality
- To map practices that coexist and co-labor together with divergence and through a negotiation of interests (ecological praxis with decolonial praxis)
- To foster divergent coalitions, to engage in shared sacrifice, to create options for inhabitation, nonmodernity, and decoloniality
- To address deficits of shared practice rather than deficits of shared knowledges

Pluriversal rhetorics in relation to ecological and decolonial rh have key methodological orientations:

marginalia: would be / clear guidelines ✱

- Ethically attend to the otherwise, the elsewhere, the diverg noncolonial/nonmodern (take into account a number of onto-epistemological worlds) on their own terms, and through engaged rhetorical scholarship
- Account for entanglement—the situated, relational, and political aspects of materiality as everyday human and more-than-human rhetorics
- Locate the researchers as embodied, situated, agents held accountable to improving the material circumstances of Indigenous and subaltern peoples, especially those from the researchers' sites of study
- Emphasize connecting one's own body and experience to the social, political, and ontological dimensions of theorizing while understanding that "other truths also exist and have the right to exist, but their visibility is reduced by the continuing power asymmetry, which is based on the coloniality of knowledge, power, being, and gender"[138]
- Movement from "a politics based on identity" to "an identity based on politics"[139]

In short, and for the purposes of this project, pluriversal rhetorics are those that study and/or practice rhetoric as an ecological (a material and relational) system that functions across heterogeneously entangled worlds against the power differentials of coloniality. San Antonio and the rest of South Texas has a long history of this colonial grating and blood soaking; a long history of extractivism that is more intense now than ever; and a long history of extreme weather events as a primary actor in these stories. Unearthing the rhetorical responses to these events from local perspectives is vitally important because, as Darrel Wanzer-Serrano reminds us, location retains a commitment to local knowledge and to a plurality of interests and perspectives that animate community control.[140] Perhaps this is one pathway toward understanding how much diversity democracy can handle, and how there can be a climate adaptation that resists assimilation and instead works toward pluralization. Yet, as the hundred-year history of climate politics this book traces will attest, our contemporary moment is not even close to considering the relations required to stop extractivism. Any form of "adaptation" will be haunted when the worst historical climatic events become everyday manifestations of colonial domination. Perhaps now more than ever, it is imperative to place the climate crisis within the context of another iteration of coloniality/modernity so ecological and decolonial scholars and activists alike may see these connections clearly, call them out loudly, and demand better futures.

PART II
Pluriversal Rhetorics

CHAPTER TWO

Rhetorical Resilience and/as Shared Sacrifice in Una Culebra de Agua

CLIMATE SCIENCE CAN NOW ESTIMATE which existing regional climate conditions are likely to get more extreme in the coming decades. San Antonio is already experiencing more days above 95 degrees than in past climate history,[1] and it will also experience more drought and flash floods due to an increased frequency in extreme precipitation events.[2] As these extremes become new normals, their effects will be felt differentially across socioeconomic lines as those with the means and mobility to adapt will do so, and those without these privileges will live even more precariously. Of course, these social locations and their associated environmental conditions are nothing new; they are products of a modern/colonial history of uneven development from global capitalist forces of production facilitated by liberal democratic governance. The introduction to this book argued that while the cumulative effects of global climate change are unprecedented, the local and regional responses to extreme weather events have a history and therefore a politics. This is what I have called climate politics—how the material forces of climate have influenced politics, and how politics have materialized understandings of climate. As unprecedented as our climatic futures may be, public and political responses to historical extreme weather events are relevant for contemporary climate politics. Just as climate scientists study past climates to understand future ones, rhetorical scholars can study the past politics of climate to understand future political capacities. These deep local histories of climate politics bring to light sticking points in our democratic cultures that will return again in a world of supercharged weather from human-induced climate change.

Because it is not possible to focus on every aspect of climate politics, the next few chapters develop pluriversal rhetorics through an examination of significant scenes from San Antonio's political history in relation to extreme flash flooding events. Within each specific event, I analyze multiple public rhetorics that respond to the floods from specific locations across the city, and I use this response to constitute a relevant rhetorical practice for political decision-making. For San Antonio, flash floods are simply a way of life. They are deeply embedded into the regional ethos of this city, and they have long shaped the contours of modern/colonial urban development. When copious amounts of water flow through the city, the security systems in place to protect people—dams, drainage canals, floodplain developments— materialize a history of rhetorical practices and policy choices enacted within a colonial social order that largely remains in place today. Yet in making the city's protective infrastructure radically visible, extreme weather events represent moments of deep introspection and political possibility. In this sense, extreme weather events create their own cosmos, or a world wherein the composition of multiple, divergent worlds and the emergent politics of which they are capable becomes differently possible. This is why the local effects of extreme weather cannot be characterized as a universal experience. Rather, these climate effects are better characterized through a pluriverse— multiple worlds co-existing together through extreme weather events and under the power differentials of coloniality.

This chapter examines the most extreme flash flood to ever hit San Antonio in order to theorize a rhetorical and pluriversal understanding of resilience. Popularly understood, resilience often marks an individual or community's ability to cope, re-center, bounce back, and return to a stable state through adversities. But a rhetorical understanding of resilience emphasizes its material and relational capacities that may not transform colonial social orders but tactically reinvent ways of living and responding to oppressive circumstances.[3] A pluriversal understanding of resilience attends to the many divergent worlds responding to extreme weather events, not for the sake of cultural comparison but rather for an analysis of climatic response-abilities that emerge within their specificity and with other responses across the city. So, while many publics demonstrate resilience in their response to the flooding event, what constitutes resilience is an un/commons—an excess of interests, practices, and worlds that are not the same, even as they emerge together through catastrophe.[4]

Amid a proliferation of discourses and practices that characterize extreme weather events as "natural disasters," rhetorical and pluriversal notions of resilience attempt to reckon with the historical and political forces of coloniality that continually place working people and people of color in precarious living conditions. In this way, pluriversal rhetorics can productively historicize contemporary concerns about climate justice. While it is important not to read historical events anachronistically, the unjust social and environmental conditions that many communities of color live through have their origins in state-sponsored colonial violence. Attending to these histories, then, is an attempt to reckon with the tragedies and the politics of historical climate events in ways that can transform mourning and loss into practices of resistance, or what Américo Paredes describes as "anamnesis," a praxis against forgetting.[5] In diagnosing the rhetorics and politics from past extreme weather events, it becomes possible to achieve some form of reconciliation while learning from our histories and working against futures of inevitable and perpetual injustice.

In order to demonstrate how this pluriversal orientation to rhetorical analysis offers insight into the political potentials of resilience through extreme weather events, this chapter proceeds in three steps. First, I situate rhetorical resilience as a capacity for publicly reckoning with the injustices of coloniality. Then, through San Antonio's local histories, I engage in a rhetorical analysis of multiple publics responding to an extreme flooding event in order to theorize rhetorical resilience as a praxis of shared sacrifice. Finally, I end by addressing the city's political decisions that reified coloniality through consequential flood infrastructure investments while also pointing to marginalized publics who used the event to seek political transformation. Throughout, I argue that the discourses and practices of resilience as shared sacrifice are a rhetorical un/commons capable of holding a city's inhabitants and governments accountable to its democratic ideals through catastrophe.

A RHETORICAL RESILIENCE TO RECKON WITH COLONIAL VIOLENCE

Rhetorical understandings of resilience have emphasized its contextual, relational, and social capacities in contrast to its more popular understandings that amplify the recovery of the self-sufficient and heroic individual or community. For example, rhetorical scholars Elizabeth Flynn, Patricia Sotirin, and Ann Brady characterize resilience as a form of tactical agency within oppressive circumstances that "mobilizes the power of imagination and reflexive meaning making in order to continually reinvent selves and possibilities and to precipitate change." For these feminist rhetorical scholars, resilience operates within oppressive and largely immutable circumstances in order to refashion identity and possibility and to create shared meaning.[6] Like rhetoric itself, resilience is a form of responsivity that can transform vulnerability into persistence, which opens possibilities for reinventing ways of living.[7] Similarly, in their study of the 2005 London subway bombings, Hamilton Bean, Lisa Keränen, and Margaret Durfy understand vulnerability and resilience as a dialectic, since one must be vulnerable and survive to demonstrate resilience. For them, resilience is a constitutive rhetoric that reaffirms a cosmopolitan national identity, partially for the purpose of anti-terror security policies.[8] Yet they also note how defining citizens as resilient can foreclose other possible responses to disasters, since resilience can also delegitimize marginal voices and circumvent public participation in national security affairs.[9]

These rhetorical understandings of resilience have been extended by scholars invested into new materialist rhetorics and their political dynamics. For example, Nathan Stormer and Bridie McGreavy argue for an understanding of resilience as a capacity to persist through systemic adaptability and sustainability, rather than individuated abilities to resist. For them, rhetorical resilience is thoroughly ecological in that it is a performance of addressivity that acknowledges diverse nonhuman relations within particular arrangements of the material and discursive.[10] Such arrangements of addressivity are also thoroughly political because they establish the potential and limits of who or what is affected by any given situation like a flooding event. Pluriversal rhetorics highlight this emphasis on arrangement and capacity as a kind of multidimensional unfolding of both matter and meaning-making across differently entangled worlds. So, while popular forms of resilience are often articulated as control and a desire to reduce vulnerability (as if that were even possible), ecological understandings of resilience are based in interdependence, where vulnerability is a potential strength and a capacity for adaptation that itself becomes a mode of persistence.[11] In asking how resilience might be done differently through rhetoric, McGreavy suggests vulnerability should be understood as a form of mutualism and a source of strength for mutual growth and transformation.[12] In this sense, resilience is a relational dynamic that animates capacities for persistence and change. Thus, the political potentials of doing resilience differently are shaped by material locations, community response-abilities, and the narratives already in place.

Yet if resilience is the capacity for adaptive persistence that relies on mutual vulnerabilities, then some distinguishing among ratios of vulnerability seems appropriate in order to discuss rhetorics and politics of resilience. After all, if taken to extremes, vulnerability can reach the limits of a capacity to be affected and can create resentment and anger from marginalized communities who are perpetually made vulnerable, sometimes to the violent edges of annihilation. Under colonial social orders, mutual vulnerability can be difficult to come by. And in cases where vulnerability becomes a continuous experience of violence, it can be irresponsible to address communities experiencing injustice as resilient. To address this power differential, rhetorical scholars have forwarded notions of precarity, which cultivates a deliberate attention to vulnerabilities as the inequities of differential exposures to violence and death.[13] It is worth noting how resilience and precarity might be mutually informative. While resilience may lack a critical attention to the power differentials of coloniality, it does theorize change through distributed forms of agency with adaptive capacities. On the other hand, precarity and its theory of change and agency is a politics of solidarity through forms of resistance.[14] Without a materialist theory of change and agency, precarity may unnecessarily frame vulnerability as a permanence, yet, without a way to adjudicate equity, resilience too easily reinforces social norms. In these ways, precarity and resilience might be co-productive in ways consequential for rhetorical studies.

So, what seems necessary for an appropriately political notion of rhetorical resilience is a sense of its adaptive persistence and its forms of solidarity that are capable of reckoning with the legacies of colonial violence. One potential way forward for this understanding of rhetorical resilience is an ecological and decolonial (pluriversal) analysis of environmental privileges—"the taken-for-granted structures, practices, and ideologies that give a social group a disproportionately high level of access to environmental benefits."[15] As Claudia Anguiano, Tema Milstein, and Iliana De Larkin and their colleagues argue, by taking environmental privilege into account, deliberations and policy decisions around resilience in the experience of extreme weather events become usefully problematized for the purpose of potential rearticulations.[16] Since arrangements of addressivity are also shaped by politics, a pluriversal understanding of environmental privileges may help trouble normative resilience rhetorics and their associated political decisions. These pluriversal notions of rhetorical resilience will only become more important in a world of climate breakdown. As Kari Norgaard notes in her ethnographic studies of climate change in Norway, "Privileged people around the world will be faced with more and more opportunities either to develop a moral imagination and imagine the reality of what is happening or to construct their own innocence from the resources of their culture's particular tool kit."[17] A pluriversal orientation to rhetorical resilience, then, allows scholars to address questions such as: how can resilience be understood as an inventive and imaginative practice that disrupts overly privileged arrangements of addressivity (many avenues for systemic adaptive persistence versus only a few, or none)? How can resilience as adaptive persistence be folded differently to reckon with colonial social and environmental orders and to foster political solidarity by reimaging relations among worlds?

To demonstrate these pluriversal political potentials, this chapter traces the commonplaces and un/expected turns of resilience through the public discourses and political practices surrounding a historical extreme weather event. More specifically, I use pluriversal rhetorics to analyze how the most devastating flooding event to ever hit San Antonio, Texas, forced a reconsideration of relations among worlds and opened up the political potentials of resilience as a version of shared sacrifice. Not coincidentally, this was the same year and place that gave rise to the first political organization for Mexican American civil rights and to San Antonio's most consequential investment in flood protection. My analysis is situated within the *kairos* of this flooding event and based on a multigenre corpus that includes multiple Spanish/English newspaper sources, scientific and engineering reports, community narratives, pamphlets, photographs, and much more.[18] I trace the un/commonplaces of resilience through these sources in order to comment on their implications for deliberation and decision-making in the attempt to transform or make intransigent inequitable regimes of environmental privilege. With an eye toward pluriversal forms of resilience, I end with a brief commentary on where crucial moments of public discourse and practice fold and how they might have been folded differently. The upshot of such scholarship is to identify a few rhetorical practices of resilience useful for reckoning with the histories and futures of climate events under colonial social orders.

CLIMATES OF INJUSTICE IN THE TEXAS MODERN, SAN ANTONIO, 1913–1925

Early twentieth-century San Antonio represents a highly fluid and modernizing cultural crossroads at the end of an era of US imperialism that transformed South Texas into an agricultural hub where ethnic Mexicana/os were largely deemed enemies of the state.[19] With a broad agricultural economy, extensive military facilities, and a railroad, investment in urban infrastructure at this time was explosive: street improvements, hotels, light rail lines, and real estate were all responsible for making San Antonio Texas's most populous city by the early 1920s with over 160,000 residents.[20] In addition, San Antonio had the largest number of people of Mexican descent in the United States: "From 1900 to 1930, the city's Mexicano population increased from 13,722 to 82,373 residents . . . in 1900 they represented 25.7% of the population; by 1930, they represented 35.7% of the population."[21] While nothing like the majority status that Tejana/os held in San Antonio prior to the 1835 revolution, and nothing like the majority status that Latina/os hold today, the early Texas modern represented an explosion of emergent transcultural publics in the context of extreme racial segregation.[22]

For people of Mexican descent in San Antonio, the end of the progressive era was the height of anti-Mexican violence in the region's checkered colonial history. Moises Sandoval, the chronicler of the League of United Latin American Citizens (LULAC), a founding organization for Mexican American civil rights, wrote that by some accounts lynchings were higher in number in the Southwest between 1865 and 1920 than in other parts of the South. No jury along the border would convict a European-Anglo for killing a "Mexican." Thus, many Texas rangers brutally assaulted and murdered Mexicana/os regardless of class or creed, and often without consequence. For Mexico-Texanos (Tejana/os), immigrants from Mexico, and Mexican nationals seeking refuge from the Mexican Revolution, what they found in San Antonio was an environment of segregation and humiliation at all levels of public life, including schools, restaurants, theaters, barbershops, and more. Yet despite the widespread violence and prejudice, neighborhood sanctuaries for those with Mexican heritage could be found in many towns and cities in South Texas, particularly in Westside San Antonio.[23]

European-American immigration into San Antonio reached its height from the period after the war for Texas Independence (1835) up through statehood (1845) and into the Texas modern (early twentieth century) with a majority of the Euro-American population coming from the southern United States.[24] This cultural crossroads was thoroughly mediated by the large number of newspapers circulating in the city. By some accounts, there were twenty-nine newspapers published in San Antonio between 1880 and 1959.[25] From the Euro-American community—itself a mix of cultures, languages, and national interests—the major newspapers consisted of the *San Antonio Light*, which by the 1920s was liberal Democratic in its political views and circulated roughly 11,000 to 25,000 copies daily,[26] and the *San Antonio Express*. The *Express* had the longest history of continuous publication in the city, beginning in 1865 and still running today. By the early 1920s, the *Express* had a circulation around 34,000. The same company began publishing the *San Antonio Evening News* in 1918, before the two eventually merged in the 1980s.[27]

For people of Mexican descent, this period offered two distinct visions of cultural identification that were complexly hybridized and mediated through the city's Spanish-language newspapers. As Richard Buitron has written, there was an elite class of Mexican professionals and nationals for whom San Antonio was a place of refuge for a political and literary agenda until they could return to Mexico. "In the words of Mario Garcia, 'they possessed a Mexican dream, not an American dream.'"[28] The elite status of these political and literary professionals is perhaps best noted by the fact that Francisco I. Madero took refuge from the Mexican Revolution in San Antonio, where he issued his Plan de San Luis Potosí, a document that helped him win the thirty-third presidency of Mexico.[29] The second group represented a Texas-born Mexico-Texano middle class, some of whom had recently returned from serving in World War I. This professional Mexico-Texano class was angered and embarrassed by segregation, discrimination, and racial prejudice from those in the Euro-American Texan community. This Mexico-Texano public sought improvement in their lives through an American dream.[30] In between these distinct groups were thoroughly hybridized publics that shared common interests in Mexican cultural identity. Thus, for those communities living in San Antonio, struggles for Mexico's independence were not easily separated from confronting Euro-American racial segregation and discrimination.[31]

With their own stores, schools, churches, and clubs, the diasporic community of Mexicana/o nationals developed a coherent cultural identity by publicizing the culture and politics of Mexico. From this community arose the Spanish-language daily newspaper, *La Prensa*. Started by Mexicano national Ignacio Lozano, *La Prensa* featured "journalistic descriptions and editorials on the events in Mexico and World War I . . . [and] was an important intellectual vehicle of expatriates living in the United States."[32] *La Prensa* is widely regarded as one of the best-written Spanish-language newspapers of the time. Local happenings normally took a back stage to the newspaper's focus on important international events from Mexico and beyond. From their outpost in San Antonio, Lozano and his writers actively created a transnational connectivity that sought a return to their place in elite Mexican society.[33] *La Prensa* helped create politically aware publics of Mexican-origin peoples in San Antonio, and these publics also became a force for the political and social activism in the mid-twentieth century of the city.[34] One of the rival Spanish-language newspapers, *La Epoca*, ran roughly from 1913 to 1931. Managed by José Quiroga, *La Epoca* was "dedicated to serving Mexican immigrants living in the United States. Its goal was to create solidarity among *la colonia mexicana* in San Antonio."[35] Both of these publications covered the flood of 1921 in San Antonio extensively. Thus, these publications helped create and serve literate, bilingual, and middle-class Mexico-Texano publics who were coming into a transnational political consciousness that would have important implications for their social activism in the United States.

These San Antonio publics were situated in one of the most flash-flood-prone regions in the United States, and despite years of deliberation, the city remained indecisive on the best way to provide security for its inhabitants. Of course, these histories of underdevelopment matter because along with the landscapes and waterways, developments co-produce a capacity for public response-abilities through their material arrangements. The early twentieth century witnessed two extensive floods in 1913 and 1914. Then, when the San Antonio River overflowed into the city in both 1918 and 1919, city officials were desperate for a solution and hired a top engineering firm from Boston for advice. While offering a number of technical suggestions, this Boston firm emphasized the exigence that flood risk presents: "We doubt if the citizens realize the ruinous loss which would result today, with the present condition of the river channels, from such a flood as that of a century ago . . . [but] a very great flood ought to be expected in the near future. . . . This disastrous flood is just as likely to occur next year as at any other time."[36] Little did anyone realize the prescience of this statement. The summer of 1921 in San Antonio was dry for two months, but when a storm broke over the first full weekend in September, the watersheds north of downtown had received twice as much rain as the city itself. By 9:00 p.m. on Friday night the San Antonio River was rising one foot every five minutes until all the rivers and creeks crested around 2:00 a.m.[37] At its height, the floodwaters from the three major rivers were six- to twelve-feet deep and rushing through the heart of the city while most people were fast asleep.

What happened over the next seventy-two hours was a dramatic disturbance that killed many more people than the official count of seventy-five and constituted the single most devastating extreme weather event in San Antonio's history. Because of San Antonio's already existing uneven development, the 1921 flood revealed much about structural inequalities during a time of rapid urban metamorphosis, and racial demarcations highlight these inequities: Of the seventy-five people who were counted as dead, all but four were living in the Mexican Quarter along the San Pedro and Alazán creek systems (figure 2). Some people went missing and some were never counted. The floodwaters were laden with fuel oil from industries north on the watershed, and automobiles, furniture, yard waste, and household goods piled up high behind the city's downtown bridges. While the entire city was underprepared for such an event, and everyone was affected, the city had also deliberately structured its security regimes according to emergent racial formations based in the inequities of colonial social orders. Thus, those already living downstream in the low-lying floodplains had little capacity for adaptive persistence, little ability to simply move to higher ground. Thus, the material arrangements of uneven development also distributed unevenly the capacity for response-abilities, and the flood made this dramatically apparent.

San Antonio's ensuing deliberations about this event did not happen in a vacuum. Rather, they were necessarily dependent on local environmental conditions, existing public discursive practice, their media instantiations, and the political infrastructure that allows communities to share in dialogue and measured action. Historian Char Miller correctly argues that the flood event of 1921 didn't transform Euro-American hegemony in San Antonio. The final decision to invest in a single dam, the Olmos Dam, was a "disturbing and remarkably skewed distribution of public benefits in one of America's poorest big cities . . . [because] it was all about protecting business's property."[38] Miller reads these actions as a failure of deliberation and decision-making. For him, the discrepancy in investment and technology meant that "those who had died in the 1921 flood had died in vain."[39] This

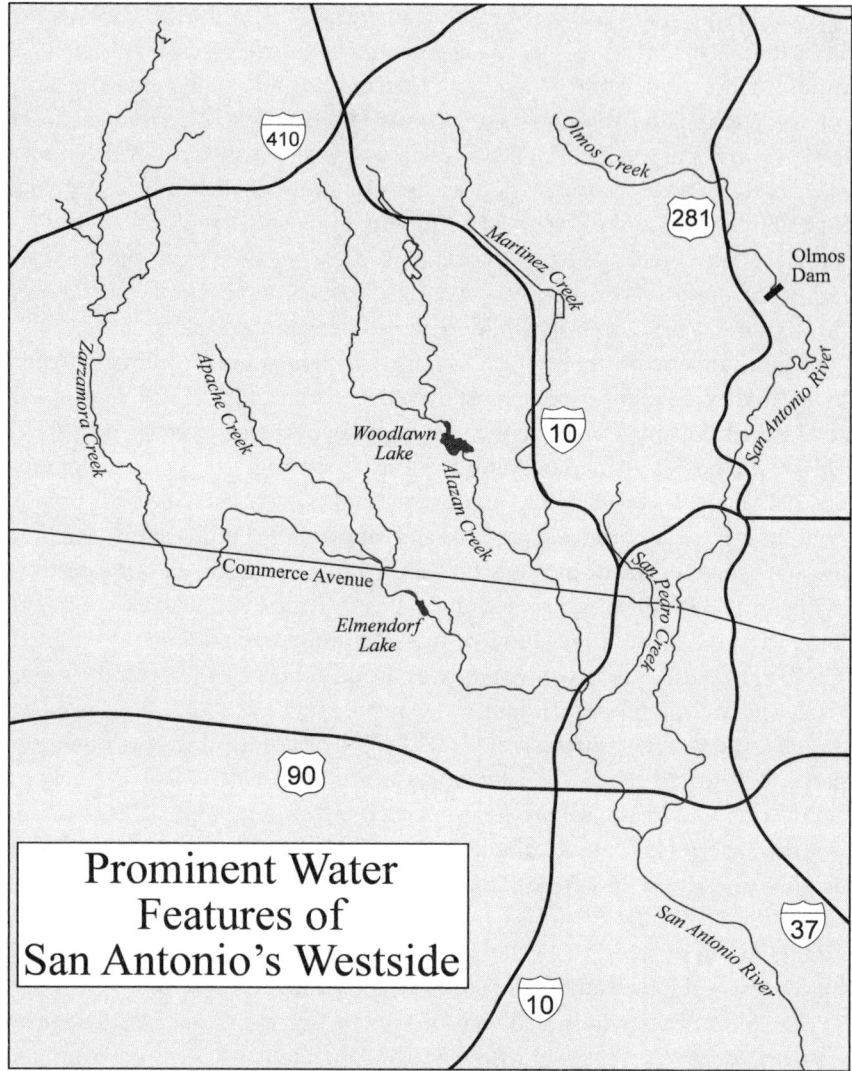

Figure 2. Prominent water features of San Antonio's Westside. Map by Charles D. Grear.

reading, however, might miss the broader relationship between the 1921 flood event, histories of racial violence, and climate politics in San Antonio generally. With an eye on environmental privileges, vulnerabilities as systemic adaptive persistence, and the possibilities of mutual transformation, my analysis of resilience during this event traces how dominant publics continually disavow their response-abilities and distort histories to suit their purposes, while subordinate publics attempt to hold their fellow San Antonians accountable through resilience rhetorics that reckon with racialized violence and offer alternative visions of political belonging. To highlight the names and stories of those who struggled against these forms of violence is to study a rhetorical form of resilience that, while perhaps not structurally transformative, continually reinvents the possibilities to redress oppressive circumstances and (re)create capacities for public and political belonging. Such rhetorical resilience is not in vain, but is another chapter in an ongoing quest to reckon with the legacy of state-sanctioned anti-Mexican violence from the situated knowledges of place and location. The next two sections take up these entangled public rhetorics directly, and then I conclude with an examination of their political effects.

THE SONG OF CONQUEST: PRIVILEGE, RACIALIZED PRECARITY, AND COLONIAL RESILIENCE

The Euro-American public response to the 1921 flood event tells a story about how discourses and practices of colonial resilience are warranted by unquestioned environmental privileges, mischaracterizations of racialized precarity, and a recentering of colonial social orders in the name of quickly reviving economic prosperity. These commonplaces unfolded through the three major English-written newspapers in San Antonio in the early 1920s: the *San Antonio Evening News*, the *San Antonio Express News*, and the *San Antonio Light*. Of course, in reports about the flood there are descriptions of the devastation, profiles of heroic deeds, responses from political figures, all with the standard tropes of resilience that call on the community to rebound and rebuild. These topologies unfold in ways that privilege the investments of Euro-American readers and mobilize victimization for political efficiency.

The strong links between vulnerability and resilience to promote political efficiencies are clearly evident in both the *Evening News* and the *Express News* where there is an overwhelming emphasis on the effects of the flood in the vulnerable downtown business district. Headlines catalog the losses in dollar amounts, and articles profile numerous affected Euro-American businesses. Nearly all of these articles emphasize that recovery for these businesses must be done expediently. When these sources address the human devastation of the flood, they recognize that all but four deaths were located in the Mexicana/o barrios. The *Light*'s coverage contains sympathetic profiles of families who suffered most, especially when rescuers encountered overwhelming barriers like piled debris and rivers merging to a width of five hundred yards. Because of these barriers, police and rescue workers heard cries from drowning people but they could not give relief.[40] None of these publications have the same sense of political expediency for ameliorating

Figure 3. Headlines from a few of San Antonio's newspapers in 1921.

the effects on ethnic Mexicans as they do for downtown businesses. These privileges afforded to downtown businesses are clearly meant to reestablish previous socioeconomic orders, but there is little attempt to question how and why people of Mexican descent found themselves in the most vulnerable positions (figure 3).

Yet because extreme weather events destabilize normalized practices, they create opportunities for engaging in cross-cultural interaction and potential mutual transformation that might not have happened otherwise in segregated cities. In this case, the deeds range from teams of swimmers that saved mothers and children from drowning to death, to distant and sluggish sympathy for the poor, to the distortions of racialized precarity.[41] The distortions of racialized precarity in particular are most prevalent in discourses of settler colonial conquest, and a then-emergent racialized theme of "dirty Mexicans."[42] For example, the conservative *Express News* column, "Mexicans Lose All on the Alazán," characterizes the victims as dirty ("covered with the slimy mud over which there was a film of fuel oil") unpatriotic (they are like "noncombatants in war"), and lazy in their eagerness to sleep.[43] For the journalist, their condition is presumed inevitable because of their social and geographical location: "Surely they could not buy a place to build on higher ground, and then would it not be years and years before another such calamity should come to pass?"[44] The writer seems not to notice his own position and assumptions, but from today's vantage point it is not difficult to see how misrecognitions reveal exactly how "Mexicans Lose All" when such events are stages to publicly reassert colonial social orders based on white supremacy, economic hierarchy, and environmental privilege.

Editorials from these publications also continually re-center and reify founding Euro-American colonial myths in order to stabilize the normative practices under threat by the flood and to assert political expediency. These resilience rhetorics are clearly connected to an individual ability to cope, resist change, reduce vulnerability, and thus regain a perception of power and control. One such passage acknowledges a direct teleology from the work of conquest to cleanliness, financial prosperity, and progress:

> The song of CONQUEST is in my heart. I personify and typify the characteristic, that, no matter how cruel the elements have been, "I WILL FIGHT ALL THE HARDER to make SAN ANTONIO THE FLOWER OF THE SOUTHWEST, TO BLOOM COURAGEOUSLY AND DEFIANTLY ABOVE THE MUCK, THE MUD, THE WRECKAGE and soon find a rooting place, in the GARDEN OF PEACE, HEALTH, HAPPINESS, BUSINESS SUCCESS AND FLOURISHING PROGRESS."[45]

The dichotomizing between mud, muck, wreckage, and peace, business success, and progress pivot on the regional identity of San Antonio as a site where colonial conquest is a constitutive element of the regional ethos. Conquest constitutes this vision of progress. Here resilience functions more like recalcitrance in its antagonistic relationship to its watersheds, and its articulation of human exception as defiance against climatic instability.[46] In the face of cruel nature, these rhetorics use the mythos of southwestern imperial conquest to make secure a particular narrative of Euro-American progress. The affectations of this vulnerability to fight all the harder for business success and progress carry political momentum toward restoring property values and the economy to their upward trajectory.

In a maneuver of intersecting colonial histories, the *San Antonio Light* contextualized the 1921 flood with a long history of flooding in San Antonio that included the early documentations of Spanish settlers who experienced an extreme flooding event during the Spanish colonial period in 1819. Unfortunately, this historicizing is used to support the false Anglo narrative that the flood had primarily come from one creek that led to downtown:

> "And all of them," stressed Mr. Smith, "were because of the flood waters that swept into the city from the Olmos creek, that stream that drains water down upon the city from the watersheds to the north of San Antonio. The old Spanish records, I recall, described the flood of 1819 as the 'Culebra de Agua,' meaning the 'Reptile of Water,' which rushed down upon San Antonio de Bexar from the north."[47]

This re-historicizing of the "Culebra de Agua" in the early twentieth century now functions in multiple ways. First, for the Spanish and Indigenous communities who lived in what is now downtown San Antonio in 1819, certainly the flooding of the San Antonio River would be of greatest concern. However, one hundred years of Euro-American conquest later, it was not the downtown river that was the main culprit of those who died in the 1921 flood event. Rather, it was the Westside Alazán and San Pedro creeks that were the main culprits for the loss of life. Thus, the county's reanimation of the snake metaphor, *una culebra de agua*, now functions topologically to falsely associate downtown watersheds with the most danger. This topology provides rhetorical force for securitizing Olmos Creek, which largely protected, and thus made prosperous, wealthy Euro-American enclaves and downtown businesses. But tropologically, figuring the creek as *una culebra de agua* highlights its metaphorical turn into a god or demon-like figure of both creation and destruction. As a figure of disturbance, the flood now also functions as a kind of colonial monster that haunts San Antonio's dominant modes of development and inhabitation. These hauntings expose the material-semiotic grammar of the relations among privileged publics that resist vulnerabilities, mischaracterize those in precarity, and re-center colonial myths to mobilize political efficiency. Among the many pathways of resilience through extreme weather events, this is a world of political and economic intransigence.

"PARA DESPERTAR EN LOS BRAZOS DEL MONSTRUO" (TO AWAKEN IN THE ARMS OF THE MONSTER): RHETORICAL RESILIENCE AS MODES OF PERSISTENCE FOR PUBLIC RECKONING

La Tragedia de la Inundacion de San Antonio: Un Recuerdo de la Terrible Catástrofe (The Tragedy of the Flood: A Memory of the Dreadful Catastrophe) (figure 4) is a sixty-two-page work of nonfiction released by José Quiroga's bookstore a few weeks after the 1921 flood.[48] The work is a collection of journalistic profiles from the flood based on interviews and fieldwork conducted by Quiroga and his team.[49] Collectively, these profiles humanize the tragedy for Spanish-reading audiences as they narrate in detail the experiences of those who endured the violence of the flood. In this way, the book unsettles the narrative found in mainstream accounts that largely trivialize, erase, or dehumanize the dead by quantifying bodies. The work is written in a highly literary style that references sources as diverse as *The Inferno*, *Macbeth*, and Cuauhtémoc—references that make it clear the intended audiences are highly educated across Euro-American and Mexican cultures. While little is known about Quiroga, like most editors of Spanish-language

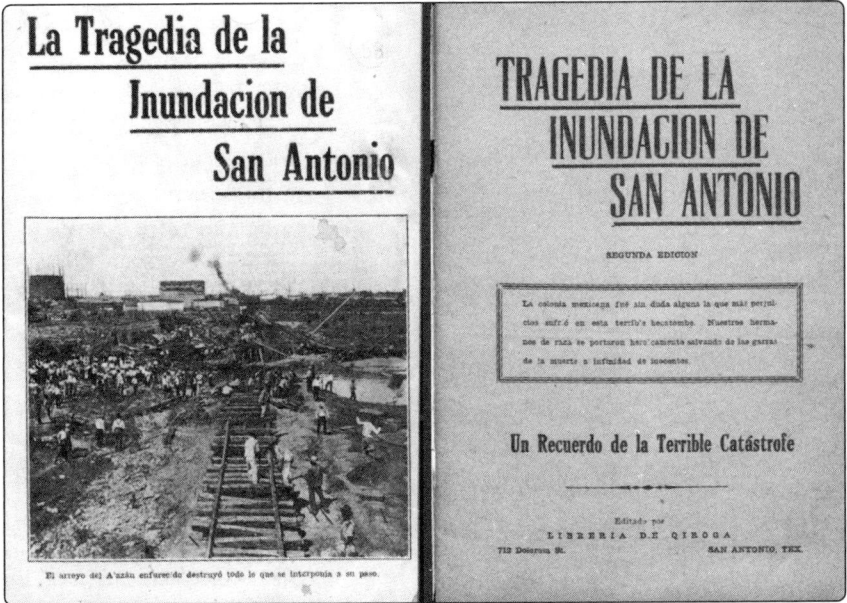

Figure 4. Pages from the book *La Tragedia de la Inundacion de San Antonio: Un Recuerdo de la Terrible Catástrofe*, edited by Libreria de Quiroga, 2nd ed. (San Antonio: Libreria de Quiroga [Qiroga], 1921).

newspapers in San Antonio at the time, he was likely a Mexican national whose philosophies of national identification with Mexico, resistance to Americanization, and rearing children with a Mexicanist consciousness clashed with the Mexico-Texano middle class who advocated for Mexicano/ as in the United States by employing American citizenship as a tool of social progress.[50] Despite these divergences, La Tragedia, along with the coverage by two other major Spanish-language newspapers, La Prensa and La Epoca, foster a cosmopolitan national identity through firsthand accounts that function both then and today as a public reckoning of the legacy of anti-Mexican violence.

Within these oppressive circumstances, Quiroga's rhetorical resilience functions complexly across cultural motives, genre, and the purpose of the speakers.[51] Portions of the speakers' purpose and cultural motives can be marked in Quiroga's epigraph, which itself is an invitation to a public reckoning of the racialized dimensions of the disaster. In the epigraph, Quiroga writes: "La colonia mexicana fue sin duda alguna la que mas perjuicios sufrió en esta terrible hecatomb. Nuestros hermanos de raza se portaron heroicamente salvando de las garras de la muerte a infinidad de inocentes." ("Without a doubt, the Mexican quarter was the most affected by this terrible disaster. Brethren from our people acted heroically, saving countless innocents from the jaws of death.")[52] Quiroga's epigraph is a recognition that while the Mexicana/o community suffered the most, they were not just victims but heroes worthy of recognition and honor. So, from the start, the themes of recognition and sacrifice are set: a recognition that the Mexican quarter suffered the most; a recognition that they suffered damages from a monstrous actor; and a recognition that they behaved heroically to save innocence. Thus, Quiroga's articulation of vulnerability and resilience are not just used as a means for the recovery of previous sociopolitical orders. Instead, they are embraced as a strength and a capacity for cultural persistence through transformations of self-sacrifice. Quiroga's memorializing of this event through the un/commonplaces of vulnerability and sacrifice folds rhetorical resilience toward a full recognition of the lack of mutual vulnerability when Mexicana/o communities continue to suffer the most damages.

As in the epigraph, a deep sense of the inequities of vulnerability reoccurs throughout La Tragedia, particularly through the unexpected figurations of the flood as a racially motivated monster. Throughout the text, the flood is described as el monstruo corbarde (the cowardly monster), acechando traidor (a stalking traitor), with insatiable jaws that eats mothers, daughters, and young lovers seeking to return to Mexico. Quiroga makes these material-semiotic figurations explicit through watershed distinctions in an early section called "Como Fue la Tragedia?" ("How Did the Tragedy

Happen?"): "Hay que ser justos, digámoslo de nuevo." ("We must be fair, let us say it again.")

> El río de San Antonio, se tragó pianos, alfombras de terciopelos, lunas venecianas de sin par primor y riqueza. (The San Antonio river swallowed pianos, velvet carpets, Venetian masks of matchless wealth.)

> El arroyo del Alazán, ahogó niños, mató mujeres, derribó hombres. (Alazán Creek drowned children, killed women, felled men.)

> Y, fué la raza nuestra, fué el pueblo mexicano, cuya escasez de recursos no le permitía habitar una casa alta, en barrio piadoso, una calle cercana al centro y fuera de peligro, la que sucumbió veneida. (And it was our people, it was the Mexican people, whose lack of resources did not allow them to inhabit a high house, a gracious neighborhood, a street near downtown and out of danger—who died, defeated.)

> Fueron los hijos de México, los que más se adurmieron, insensibles al peligro, para despertar en los brazos del monstruo. (It was the children of Mexico who slept the most, unaware of the danger, only to awaken in the arms of the monster.)

> Fueron pues los arroyos de San Pedro y del Alazán, a los que más odio debemos guarder. (But it is for the San Pedro and Alazán creeks that we should hold the most hatred.)

> Porque el rio, acabó con la riqueza, mientras que éstos, los traidores, se hartaron de carne humana, para vomitarla putrefacta, asquerosa y nauseabunda. (Because the river, it destroyed wealth, while these, these traitors, gorged on human flesh to later vomit it up, putrid, revolting, nauseating.)[53]

This passage dramatizes environments of inequitable sacrifice in the experience of extreme weather events. Landscapes and watersheds become differentially violent in relation to material goods and racialized human bodies, particularly expressed through Quiroga's verb choices: The San Antonio River se tragó (swallowed) material wealth. But the Alazán Creek ahogó niños (drowns children). The San Antonio River acabó con la riqueza (destroyed wealth), but the San Pedro and Alazán actively se hataran (gorged). It was only the Sons of Mexico who para despertar en los brazos del monstruo (awaken in the arms of the monster). In Quiroga's watershed distinctions, he identifies a colonial monster that is violent, deadly, and yet a traitor and a coward for drowning Mexicana/os in their sleep. So, how did the tragedy happen? For Quiroga, the tragedy was the lack of mutual vulnerability and the recognition that perpetually underresourced Mexicano/as were violently sacrificed by a racially motivated monster. In this way, the 1921 flood is yet another tropological verification of colonial relations of economy and environment that legitimize and perpetuate anti-Mexican violence.

In the epilogue, Quiroga and his team write that this book was meant to perpetuate these stories so as to remember them: al escribir este libro, . . . nos llevó el objecto, de perpetuar en él, la memoria de estos hechos (in writing this book . . . we were guided by the goal of perpetuating in it the memory of these events). Yet, they also state another purpose: to transform vulnerability into sacrifices as public acts of cultural persistence and self-transformation. In doing this, they open the question of what political loss and political sacrifice can mean for subordinated publics in a polity:

> Hemos llevado otro fin . . . el de demonstrar, que la raza heroica, está siempre presta al sacrificio, cuando éste puede beneficiar a la humanidad. ([We] have had another end . . . to demonstrate that our heroic people always lend themselves to sacrifice when it can benefit humanity.)[54]

> Hemos dicho también, que hemos querido exalter en él, a la sangre Hermana, al mexicano, que claudicante y doliente se arrastra y va por la vida, pobre, miserable y triste, pero llena el alma de un valor indomable, de un herocio y noble sentir, que nada ni nadie, hará torcer. Al pasar los ojos por estas escenas, se verá que son águilas, que son leones, que son guerreros siempre indomables los mexicanos. (We have also said that we wanted to extol the united blood, the Mexican, who, limping and in pain, pulls himself up and reaches for life: poor, wretched, and sad, but with a heart full of unbreakable courage, with a noble, heroic sense that nothing and no one can turn aside. As your eyes move over these scenes, you will see that they are eagles, they are lions, they are always indomitable warriors, these Mexicanos.)[55]

Understanding sacrifice from the perspective of Mexicano/a cultural practices has a long and complex history that cannot be fully captured in this analysis.[56] However, in terms of rhetorical resilience what is striking is how Quiroga transforms vulnerability into a resource when the sacrificial acts of the Mexicana/o community function as a mode of transformation and persistence. These acts of rhetorical resilience resist erasure, partially through a recognition of a larger political quest to benefit humanity, and partially through a self-transformation into eagles, lions, and indomitable warriors. Yet, the widespread misrecognitions of these acts are a threat. Again, through tropes of the human-environmental monster and its agency, Quiroga remarks that "nothing and no one can turn aside" the value and nobility of this sacrifice. So, the misrecognition of what is noble, heroic, and unbreakable does not just arise from a thing but from *a someone*. Linking back to the context of extreme segregation, for Quiroga, this extreme weather event fosters political emotion because it feels like the very landscape—a thing—has the capacity for cultural erasure. Yet, the rhetorical resilience to record these acts of violence, to make them recognizable, and to reinvent them as modes of persistence is to create a political language of shared sacrifice for the perpetuation of a transnational Mexicana/o identity. These recognitions and reinventions fulfill his purpose to extol the united blood of the Mexicana/os, as does the choice to end his narrative with the cosmopolitical national symbol of an eagle on a cactus devouring a snake.[57]

How are polities meant to deal with the problem that subordinated publics are constantly and unjustly called to make the greatest sacrifices? In other words, how do polities deal with the problem of democratic sacrifice so that vulnerability does not just become victimization and a breeding ground for political distrust?[58] As we have seen, extreme weather events dramatically highlight the structures of colonial social orders, characterized by the lack of mutual vulnerability, and the racialized dimensions of inequitable democratic sacrifice. In this sense, extreme weather events can become rituals of violence and racialized sacrifices that fail to distribute pain and render it communal. It is the lack of mutual vulnerability and inequitable sacrifice exposed by the flood that is the cowardly and traitorous monster enacting racialized violence. A public reckoning of this monster is not simply an exercise in guilt. Through their journalistic work on the flood, Quiroga and his team consciously engage in a public reckoning of anti-Mexican violence, and it's clear they do this so reading publics might recognize this monster, see it, and then honor those who were sacrificed by it. This is another fold of a material and relational form of rhetorical resilience—an honoring of those who sacrificed so the larger community can live on, and do well by those who gave their lives. Such recognition and honor then make it possible to love this monster with a decolonial sense of love—a call to engage in mutual vulnerability and to put into practice legitimate shared sacrifices as a form of solidarity.

RHETORICAL RESILIENCE AS SHARED SACRIFICE AND POLITICAL SOLIDARITY

The preceding sections traced the commonplaces and unexpected turns of resilience during this extreme weather event largely as a reassertion of cultural factionalism. Though both Anglo and Mexicana/o communities recognized the most basic of realities of the flood, the commonplaces of privilege and racial misrecognitions from Euro-American publics and those of inequitable sacrifice and cultural resilience from Mexicana/o publics largely reinforce cultural borders. Yet other forms of rhetorical resilience deserve attention, particularly those that refashion privileged social positions as a relational capacity to act together in the face of tragedy. As Quiroga surely understood, as much as the event is a tragedy characterized by inequitable and racialized violence, it is also an event characterized by acts of human compassion and solidarity that transcend the normalized borders of cultural communities.

In recording the acts of people who forego their privileged social locations, and who join to help those in need, Quiroga's *La Tragedia* demonstrates a capacity for legitimate sacrifices to create mutual vulnerability and political trust. While for some such practices may be temporary, and overly reliant on the mobility of privilege, they are also practices of mutual recognition and engagement that stand in stark contrast to the misrecognitions and defensiveness of white victimhood in the previous section. Quiroga writes of ladies emerging from homes pampered by fortune who swap their rich silk kimonos for "el humilde vestir, albeante, de las damas de la Cruz Roja" ("the humble white costume of the ladies of the Red Cross") . . . a "laborer por el bien y por la caridad" (to "labor for good and for charity").[59] The men who joined them are described as gentlemen of high ancestry, and "no van a un banquete, ni asisten a una 'soiree' van a la más grade de las festividades . . . van, en el ejercieio del bien, al luchar por la caridad y por los que han, hambre y sed de piedad." ("They were not going to a banquet, nor attending a soirée. They were going to the greatest festival of all—they are going in the exercise of good, to fight for charity and for those who hunger and thirst for piety.")[60] In reframing deliberation around acts of recognition and reciprocity, Arabella Lyon notes that deliberative acts are not a consensus but rather a practice embedded in a commitment to sustaining recognition and engagement even in fractured, fractious, transient situations.[61] Quiroga's portraits of privileged individuals who engage in a modicum of sacrifice, and who take on the responsibility to save innocence, demonstrates these acts within segregation culture. As Stengers would note, such acts are not independent of the environment and ecology from which they arise. It is the event of the flood that actualizes the constraints and obligations of rhetorical resilience in ways that cannot be completely predicted. And yet through such events one may learn about how these relationships foster creative solutions to rearticulate rhetorical resilience through legitimate sacrifices and mutual response-abilities.

These deliberative acts culminate in Quiroga's portrait of Mrs. Fogleson, whose responsibilities are celebrated as a version of reciprocity and a practice of shared sacrifice:

Una dama que siempre se ha preocupado por el pueblo mexicano. Sin estudiar español, habla ya mucho de nuestro idioma, de tanto rozarse con la clase menesterosa de nuestra Patria, radicada en esta población. (A lady who has always looked out for the Mexican people. Without studying Spanish, she already speaks much of our language from so much time spent in contact with the underprivileged of our Homeland who are settled in this city.)

En esta vez, desplegó todas sus actividades, fué de aquí para allá, no tuvo sino instantes muy cortos para descansar y tornar a la brega. (On this occasion, she deployed all of her activities, going from here to there; she only had the briefest of moments to rest and then she returned to the struggle.)

En nombre de México, en nombre de nuestro país, hay que agradecer a esta dama, su celo y su amor, por los nuestros. (In the name of Mexico, in the name of our country, we must thank this lady for her zeal and her love for our own.)

La hemos visto inclinada, enjugando cariñosa el pie enfermo y lodoso de un anciano . . . la hemos visto, tener en sus brazos un niño, herido y cuyos lamentos llegaban a el alma, manchar sus albas ropas, oprimiendo en sup echo aquel pequeño ser que se moria. (We have seen her bent over, lovingly washing the sick, muddy foot of an old man. We have seen her holding in her arms a wounded child whose cries pierced the heart, soil her pristine clothing; pressing that small, dying child to her breast.)

Hay que agradecérselo, repetimos . . . en nombre de México y nombre de sus hijos! (We repeat, she must be thanked for this—in the name of Mexico and that of her children!)[62]

What Quiroga makes clear is that Mrs. Fogleson's commitment to return to the struggle is not just a religious or occupational devotion but a political act that looks out for the underprivileged through constant contact, and a constant return to the struggle. Through these practices, Mrs. Fogleson not only recognizes the sacrifices of the Mexicana/o community but also honors them through compassionate acts of care, bilingualism, and struggle. This act is public in the sense that it cares more for the conditions of specific people and communities than political ends, which is a form of solidarity and a response-ability in moments of crisis.[63] Thus, this extreme weather event is not just characterized by inequitable sacrifice, but by a rhetorical resilience as a creative potential for mutual recognition, response-ability, and reciprocity in difficult situations. In Mrs. Fogleson, Quiroga finds solidarity across cultural differences, and in his own act of recognition, he repeats, she *must be thanked* for this.

These portraits of reciprocity demonstrate an important point about shared sacrifice that reveals the limited ability of liberal democracy to truly address racial justice. The gratitude Quiroga shows to Mrs. Fogleson is an important moment of solidarity, but it also demonstrates that there is likely never to be a true equality of sacrifice (where everyone is asked for the same kind of sacrifice). Rather, there is what Stengers would call a putting into equality of sacrificial practices, where what it means to sacrifice is differentially understood and practiced from already unequal social locations. The already existing structures of racial inequity meant that Mexicana/os were sacrificed unknowingly, while for characters like Mrs. Fogleson, sacrificial practices are a choice, an ethics of relationality, that ask one to embrace vulnerability to serve a higher purpose. In putting into equality sacrificial practices, everyone in their own ways of life and in their own social locations is called to embrace vulnerabilities as potential strengths and modes of persistence that achieve some measure of political solidarity. In other words, sacrificial practices open the capacities for mutual vulnerability to become political solidarity. Here this solidarity is limited to individual portraits while the larger colonial structures of urban life for Mexicana/os remained monstrously violent. Beyond these gestures of solidarity, there is little equity to be found because at a structural level Mexicana/o sacrifice was never truly honored, but rather misunderstood, misrecognized, and then forgotten. The victims of racialized sacrifices cannot also be expected to bear the burdens of equity and justice alone. Rather, it is those in positions of privilege who should engage in an ethics of response-ability to transform the inequities of sacrifice and to more fairly distribute pain and suffering across a polity. Honoring shared sacrifices can only ever be one step in the direction toward more resilient civic relations. The broader story of this extreme event calls into question the assumptions of racial progress and the perfectibility of US democracy, especially when the majority of privileged publics fail to recognize the inequities of sacrifice and instead respond with their own sense of defensive vicitimhood, and of course, their own version of what it means to politically redress such experiences. In this sense, Quiroga's achievements in rhetorical resilience are a recognition that the dead are not gone, and that their sacrifices still require a public reckoning if there is to be any political solidarity.

EXTREME WEATHER EVENTS AS CATALYSTS FOR POLITICAL INTRANSIGENCE AND TRANSFORMATION

The San Antonio flood of 1921 has many stories to tell, all of them with their own particular lessons about rhetorical deliberation and political capacity in the midst of extreme weather events. In studying this event, I have argued that the Euro-American publics in San Antonio were largely concerned with property rather than people. I demonstrated this by tracing the commonplaces of resilience to privilege, racialized precarity, and colonial myths. On the other hand, Mexicana/o publics, while also concerned with business and property (after all, Quiroga was also there to sell newspapers), were largely concerned with the survival of their people. I demonstrated this by tracing un/expected turns of resilience to practices of shared sacrifice as key political rhetorics for contemporary notions of mutual vulnerability, adaptive persistence, and resilience. In Quiroga's reporting in particular, he demonstrates how mutual sacrifice and democratic solidarity requires a compassionate critique of privileges, especially by publics who refuse to recognize, and yet still benefit from, the sacrifices of others.

In this case, the colonial commonplaces of conquest, racial superiority, and privilege solidified Euro-American resilience as an expedient path toward strengthening flood protections in a dramatically unequal fashion. These topologies were dominant in the public discourse of this segregated city in the 1920s, and they brought rhetorical force to the vote to build a dam across the Olmos creek that only protected downtown business and the most privileged Anglo neighborhoods. In fact, one can trace these discursive practices through the engineering reports, political deliberations, and construction of Olmos Dam where public arguments about downtown protection, rising property values, building tourist attractions, and conquering the flood peril were commonplace. But importantly, in a dramatic manifestation of the uncommon and pluriversal practices across entangled worlds, the community of scientists and engineers who first studied the flood continually called for flood protection investments on the downtown and the Westside creeks. The committee of flood protection for the city of San Antonio recommended spending $2.5 million on the northern watersheds, and $1.5 million on the Westside creeks.[64] While not entirely equitable, such investments would have created pathways toward social and political solidarity. Though city officials continued to promise a plan with protective measures for the entire city, the final decision was telling: $3 million for the construction of Olmos Dam on the north; $6,000 to widen and clear the Westside Alazán and San Pedro creeks.[65] Such monstrous disparities speak volumes about the ability of extreme weather events to reinforce segregation practices in consequential policy decisions with generational effects. As Char Miller notes, this was arguably the city's most important public works project ever—it facilitated the development of a pair of suburban enclaves that sheltered the city's Anglo elite, it provided security for capital investment into downtown skylines, and it revived the old idea of a River Walk.[66] But what must also be emphasized is that this inequitable policy decision was not supported by the technoscientific community studying the flood, whose scientific obligations to skepticism seemed to have protected them somewhat from bias. Rather, inequity was an effect of anti-Mexican political decisions.

That's one story. It tells us that in turning this flood event into a cause for thinking about the politics of climate, the active participation of all of those affected was certainly present in public, but not in the political decision to build Olmos Dam. In the ensuing deliberations around flood protection, there is a disturbing efficiency about the decision to protect downtown only. Politically speaking, protection for the Westside Mexicana/o populations was disqualified a priori and constituted yet another colonial act of anti-Mexican violence. In this sense, the flood was not a cause for political transformation but for a deepening political intransigence. Folding these policy discourses and practices differently through rhetoric would require recognition of the legitimate sacrifices of the Mexicana/o community and a political zeal that could pass equitable policies for flood protection by holding privileged people accountable to some modicum of shared sacrifice. While the voices to support equitable practices and policies are clearly present in the historical documentation, it would take another generation and a half before they would be legitimized through the city's political infrastructure.

By 1921, new visions of political belonging were not just articulated by Quiroga, but by a broader community of Mexicanos who barely a month after the flood gathered in a barbershop on the Westside and officially established the first Mexican American civil rights organization, the Orden Hijos de America (Order Sons of America, or OSA). Unlike other earlier civil organizations like *mutualistas*, the OSA was the first to organize around the idea of protecting its constituents through *political influence*.[67] Barely a year later, they drafted their Declaration of Principles and a Constitution in 1922. These constitutional documents reflect a similar demand for equity and for improving the condition of people of Mexican descent as those found in the flood reporting of *La Tragedia* and some of the scientific reporting. But for the OSA, these demands were now routed through a political organizational structure based in US citizenship and rights.[68] For example, the OSA transformed inequitable sacrifice into calls for legal protection: "to evolutionize and establish in our households the principle that we must adopt in their entirety *the standard living conditions of the American people* . . . to provide aid to our sick and distressed members, to bury the dead and to provide such *other protection* as we may be able to give our members."[69] Such calls for protection came through an embrace of the American dream of full citizenship, and all its assimilationist tendencies, as a path to secure the mutual obligations highlighted by the 1921 flood. Lacking the recognition and protections from local publics and politics, resilience as shared sacrifice here is translated into a democratic language of legal protection for standard living conditions.[70] In these ways, the public rhetorics from the flood are more verifications of sociopolitical realities that could have achieved much more protection for standard living conditions than they did. But those rhetorical practices were politically muted until groups like the OSA began to form and seek protections through citizenship.

What do these un/common places of resilience and their un/expected turns toward privilege and shared sacrifice teach us about new worldings that will only deepen the experience of these political and climatic catastrophes? If climate change is indeed about political change, then rhetorical practices that legitimize mutual vulnerability, shared sacrifice, and solidarity across different worlds will be key to appropriately politicizing climate resilience and adaptation. As crucial as climate science is for these practices, it is in public and political rhetoric that we may create negotiations across diverse and unequal social worlds exacerbated by the effects of climate change. These negotiations might also begin with recognition of how extreme weather events have facilitated inequity in the past to provide lessons for our negotiations of the future. Institutionalizing shared sacrifice through the rights and responsibilities of national citizenship, including the right to a standard of living with its attendant environmental privileges and securities, seems necessary yet insufficient. Those who died in the 1921 flood did not die in vain. Rather, they sacrificed for a larger political quest to reckon with colonial monsters and create many worlds of systemic adaptive persistence. These climatic response-abilities also constitute a heroic attempt to improve life for all in South Texas and beyond. These practices of shared sacrifice for political solidarity through catastrophic events is another story of resilience well worth remembering.

CHAPTER THREE

Divergence and Diplomacy as Pluriversal Rhetorics of Coalitional Politics

We who are colonized or function in some way otherwise cannot be the only ones leading the charge to delink from modernity/coloniality. An ethic of decolonial love requires those who benefit most from the epistemic violence of the West to renounce their privilege, give the gift of hearing, and engage in forms of praxis that can negotiate more productively the borderlands between inside and outside, in thought and in being.

DARREL WANZER-SERRANO, *The New York Young Lords*

The human environment and the natural environment deteriorate together; we cannot adequately combat environmental degradation unless we attend to causes related to human and social degradation. . . . We have to realize that a true ecological approach always becomes a social approach; it must integrate questions of justice in debates on the environment, so as to hear both the cry of the earth and the cry of the poor.

POPE FRANCIS, *Ladauto Si'*

IN JUNE 2017, OVER FOUR HUNDRED mayors representing sixty-eight million Americans formed a coalition called the US Climate Mayors. Together they organized against the withdrawal of the United States from the Paris Agreement on climate and chose to "push for new action to meet the 1.5 degrees Celsius target, and work together to create a 21st century clean energy economy."[1] In creating this alliance, the Climate Mayors movement filled an absence of national leadership by putting into action climate plans to reduce carbon emissions and adapt to the climate forces already here and coming. Such plans hope to create "climate ready," "climate resilient," and "climate adaptive" communities. When I attend climate action meetings in my adopted hometown of San Antonio, Texas, I hear many questions regarding these fuzzy terms: How can cities like San Antonio be climate resilient when carbon pollution and air quality are global and regional issues, not local ones? How can local actions muster any power when they are so profoundly shaped by state and federal law, let alone global and neoliberal economic forces? How can cities articulate together their climate actions to have any broader effects on extractivist development and privatized transportation models?

These kinds of questions reveal that the way climate comes to matter locally is nothing new. Rather, locally adapting to the climate shocks already underway engages long-standing quality of life issues around housing, energy, transportation, and more. As I noted in the introduction, the very concept of climate connects science to politics through notions of prevailing weather patterns and prevailing trends of public opinion, or political climates. I also noted how at its core, climate is profoundly regional and rhetorical—it connects place-based material conditions (environments) to public discursive practice (meaning-making) through the situated norms of creating worlds together (politics). Therefore, as rhetoricians begin to examine the local effects of climate change, one can expect to encounter familiar topics in public rhetoric that highlight theoretical problems with coalitional politics (How do they emerge? How do coalitions make meaning together? What makes them endure or fade? What gives them rhetorical power?), and theoretical problems of materialist rhetorics (How is agency a function of a distributed and more-than-human rhetorical practice? What are the political forces of materiality?). The current climate crisis may be unprecedented, but responses to extreme climate and weather events are not. Thus, histories of climate and weather responses are worth studying from local and rhetorical standpoints. If previous extreme climate and weather events are a new norm, local histories of public and political responses will be key to diagnosing the relationships among changing weather and changing politics. After all, like the weather, all politics are local.

From local standpoints, then, there can never be a single thing called a climate impact; there can only ever be many. The historical research behind these forces shows how such events amplify and reinforce previously existing social orders.[2] So, the local effects of extreme weather are not universally experienced, and they do not affect humans alone; rather, these local effects, and the responses to them, are perhaps best characterized as a pluriverse—a world of many worlds coming together through an ecology of practices to negotiate their difficulty in being together in heterogeneity.[3] In the context of anthropogenic climate breakdown, it's worth asking questions about pluriversal forms of local responses that negotiate being together through climate shocks, especially at the neighborhood level with its shared

infrastructures like public spaces, waterways, stores, restaurants, community organizations, block clubs, and churches that allow people to gather and
keep tabs on one another during a crisis event.[4] Addressing climate politics
at the local level today, then, means using rhetorical theory and practice as
a means of publicly reckoning with the legacies of colonial social orders in
order to articulate what quality of life might mean together, and in plurality,
when living through climate breakdown.[5]

Because climate politics are a pluriverse, I have argued for the necessity
of a rhetorical heuristic and hermeneutic to address the political capacities
of relations among worlds that negotiate their difficult being together in heterogeneity.[6] But these inventional and interpretive modes—the topologies
and tropologies so productive for ecological rhetorical theory and practice—
cannot just be based in citizenship or public and political actors but rather
in actively considering the ways in which rivers and creeks, for example, participate in the process, at the very least through their entanglement and engagement with a world and for a world.[7] Traditionally, rhetoricians may understand such political work through the notion of coalitions, the visions
and practices that are oriented toward engaging others through a shared
commitment to social and political change.[8] But the pluriverse challenges
a human-only interpretation of coalitional work and encourages coalitions
to consider the specific material-semiotic quality of human/nonhuman relations. As Adela Licona and Karma Chávez have written, coalitional politics
are most effective when they address oppression and power at its roots and
use difference as a resource to build more livable worlds in relation to others
whose lives, interests, and material conditions are different from their own,
including differences across human/nonhuman relations.[9]

To extend ecological and de/colonial views of coalitional politics through
pluriversal rhetorics, this chapter offers the un/common places of divergence and diplomacy as practices useful for understanding the emergence
of coalitional politics for communities responding to extreme weather
events. In particular, I extend rhetorical thinking on Stengers's ecology of
practices and focus on her use of divergence and diplomacy as matters of
arrangement, which she suggests are matters of bordering, and therefore,
by an extension I draw here, matters of border thinking and pluriversal
rhetorics.[10] As we have seen, border thinking arises from the point of colonial contact and entails the disruption of dichotomies by bringing subalternized discourses and practices into articulation with dominant logics,
and through border thinking, the pluriversal attempts to create a world of
many worlds.[11] Divergence and diplomacy share a close affinity with critical border thinking and pluriversal praxis because they represent an internal struggle against coloniality/modernity through an examination of its

epistemic borders and its neglected ontological differences—the outsider within. In the performative articulation of contact through difference, there is always divergence—sites of partial connection that arrange bodies, discourses, and materialities of space for an achievement also characterized by its excess.[12]

To demonstrate this pluriversal approach to coalitional politics, I offer a case study from San Antonio during the 1970s that fostered a coalition between a second-wave Chicana/o community organization called Communities Organized for Public Services (COPS), and a local environmental organization, the Aquifer Protection Association (APA). By thinking with extreme flash floods, COPS learned to fight against uneven development, and by thinking with the aquifer, the APA fought to protect clean water against suburban sprawl in recharge zones. The COPS/APA coalition that came together for a few years helped rearrange San Antonio's city politics toward pluralism, and the coalition's story speaks to the emergent properties of divergence and diplomacy when articulated to broader political and climatic forces. While rhetoric is often used as a key analytic for social movements, community organizations like COPS and APA exceed the boundaries of rhetorical theory, and this excess provides a useful resource for retheorizing rhetoric through pluriversal relations among worlds.

This chapter proceeds first by deepening this book's examination of the un/common places of divergence and diplomacy through the nonmodern philosophical work of Isabelle Stengers and others. As we will see, divergence and diplomacy through an ecology of practices is particularly apt for theorizing rhetoric and materiality in coalitional politics. Then, by deepening this project's examination of de-linking and border thinking through the works of Walter Mignolo and J. David Cisneros, I demonstrate some of the overlap among nonmodern and decolonial praxis, while also noting lingering tensions, such as the former's inattention to colonial race relations and the latter's de-emphasis on human/nonhuman relations and the political force of materiality. Finally, I use these rhetorically inflected understandings of divergence and diplomacy to analyze a case study of coalitional politics that helped transform San Antonio's city politics toward a pluralistic social order. Such a study offers rhetorical lessons for exactly how climate adaptation practices can "climate proof" or "climate ready" its neighborhoods in ways that address the legacies of colonial social orders.

DIVERGENCE AND DIPLOMACY FOR
PLURIVERSAL COALITIONAL POLITICS

Isabelle Stengers's ecology of practices is a nonmodern approach to ontological pluralism that methodologically insists on marking differences and "thinking in the presence of," and thus, it can be usefully articulated to pluriversal notions of coalitional politics. For Stengers, ecology captures the ways in which practices are defined by relationships, by the association between the ethos of a practice and its *oikos*, "not only the matter-of-fact environment but the way it defines its relation with other practices and the opportunities of the environment."[13] Because an ethos can only be defined in relation to its *oikos*, changes in environments are changes in relations and events that transform practices. And for Stengers, the term *practice* is not descriptive but speculative. Practice implies a situated belonging where ecological relations constrain and oblige practitioners in ways that cannot be predicted, yet through an ecology of practices we can learn about these relationships and their possible creations. For example, both of the community organizations I describe in the next section arose in response to San Antonio's suburban development, which threatened the city's source of clean water and furthered urban core disinvestment policies. These movements were not necessarily predictable. They arose from situated belongings to a changing city *oikos* that obliged and creatively constrained a new definition of the city's ethos. These dynamics, in turn, produced a coalition that fought against suburban development and created a local capacity for rearranging the city's political infrastructure. The speculative nature of an ecology of practices marks the creative capacities of Stengers's notion of a cosmos—not a single universal ideal order but rather that which is "constituted by multiple, divergent worlds and to the articulations of which they can be capable."[14] Crucially, these articulations are cosmic events, "a mutation which does not depend on humans only, but on humans as belonging, which means they are obliged and exposed by their obligations."[15] Such points are important because they remind us that politics are not just acts of collective will; politics require moments or events that oblige coalitions.[16]

Divergence and diplomacy are two related terms from an ecology of practices that can provide some direction for pluriversal rhetorical work, and it is a direction that speaks directly to sites of coalitional politics. Like Haraway and other feminist science studies scholars, Stengers exemplifies that thinking with materiality cannot supersede thinking with historically marginalized peoples with their own configurations of materiality and meaning, their own situated knowledges. For example, when Stengers asks the question of how to turn a river into a cause for thinking, she writes that it requires attention to the emergence of diverging minorities where politics has to proceed "in the presence of" those who would be most affected by the decisions, and they cannot a priori be thought of as disqualified and unable to contribute to a common account.[17] Additionally, turning the flood into a cause for political thinking is a question of material-semiotic events, or as I argue later, a question of coalitional moments.[18]

Stengers also asks the question of how to design the political scene in a way that actively protects it from the betrayals of liberal humanism, and that question can be directly addressed to rhetorical scholars who take up the notion of diplomacy. Stengers equates diplomacy with speaking about borders as matters of arrangement: "Borders do not mean that connections are cut but that they are matters of arrangement . . . with different risks and challenges for each involved party."[19] Diplomacy is not just skillful negotiation, but "an art of artificial arrangements . . . [whose achievements are] the event of an articulation between protagonists constrained by diverging attachments and obligations in situations where contradiction seems to rule."[20] Diplomacy as an art of bordering, as matters of arrangement, responds to the question of how to design the political scene in a way that actively protects it from the fiction that humans of good will decide in the name of the general interest. By attending to the arrangement of material conditions, what matters is a collective becoming that humans could not produce "'by themselves' but only because of the situation that generated the power to make them think."[21] The river and the water participate at the very least through their entanglements and engagements. Any rhetorical notion of diplomacy then cannot be thought through without tracing the arrangements among bodies, discourses, and the materialities of space through shared acts that attend to diverging minorities.[22]

Nonmodern scholars have recognized that diplomacy is a rhetorical exercise about composing common worlds through difference and human/nonhuman relations under shared concerns.[23] What Stengers adds to this conversation on diplomacy is precisely divergence wherein shared matters of concern are only ever sites of partial connection that also perform otherwise. Similarly, rhetorical scholars have associated the art of diplomacy with sophistic rhetoric wherein the composition of a political collective is based on "the practical work of negotiation facing humans and nonhumans in the uncertain flux of unfolding events."[24] In his summary of Bruno Latour's project AiME, Nathaniel Rivers notes how "diplomacy cautiously and patiently composes a common world fully attuned to the heterogeneity of multiple modes of existence." In this sense, "diplomacy is not a negotiation among different perspectives on the same underlying reality; it is the precarious and unending work of traveling to and from multiply nested but nevertheless distinct realities: something itinerant, 'passing' sophists are historically predisposed to do."[25] As we will see in the example of Ernesto Cortés and COPS, divergence within diplomacy captures this traveling work among multiply nested distinct realities by exceeding the boundaries of sophism in a way that provides a resource for rethinking diplomacy through pluriversal praxis. In the words of María Antonietta Berriozábal, the first Latina to ever hold a city council office in any city in the United States, at the time, people like Ernesto Cortés were leading multiple lives.[26]

But what happens to Stengers's theses of divergence and diplomacy if they are approached otherwise? If indeed diplomacy is an art of bordering through material-semiotic arrangements, then certainly they are differently produced when practicing from the exteriors of modern/colonial power characterized by differential movement and an ethic of decolonial love.[27] Hints toward an answer are partially provided by the work of Marisol de la Cadena, who adopts ecology of practices to analyze the mutual entanglements and partial connections of Indigenous (Quechua) and non-Indigenous worlds in Peru. For her, divergence and the art of diplomacy are creative sites of connection among heterogeneities that enable nonmodern analyses with particular attention to radical differences.[28] Referencing Stengers, she writes, "Divergence does not presuppose homogeneous terms—instead, divergence refers to the coming together of heterogeneous practices that will become other than what they were, while continuing to be the same—they become self-different. Thus conceptualized, the site where heterogeneous practices connect is also the site of their divergence, their becoming with what they are not without becoming what they are not."[29] Divergence in the art of diplomacy is an analytic that attends to excess as a way to resist and contaminate a coloniality of politics that proposes to build one world through cultural assimilation. Partial connections, "the becoming in divergence with," acknowledges mutual entanglement within ontological pluralism. Or as de la Cadena writes, "The 'other' is always part of them, as much as they are part of it. This is the partial connection that neither modern politics nor indigeneity escape: they are entangled in it, exceeding each other in mutual radical difference."[30] For de la Cadena, divergence is a tool that marks colonial difference while also marking the site of partial connection, or what she calls a decolonial practice of politics "with no other guarantee than the absence of ontological sameness."[31] In this way, divergence in diplomacy is a tool for presenting colonial difference that is still within colonial commands but divergent because something always escapes, and someone always prefers not to heed those commands. Adopting terms from European thinkers is potentially fraught with the reproduction of colonial violence, yet it is Stengers's and de la Cadena's insistence to meet marginalized discourses on their own terms that helps resist this reproduction methodologically by speaking in the presence of and attending to diverging minorities who would prefer not to.

The insistence to meet marginalized discourses on their own terms and at sites of partial connection is the exigence for divergence and diplomacy as a pluriversal rhetorical praxis. Such work is radically immanent, situated, and rhetorical through invention and guided interpretation—the topologies and tropologies that build on un/common places and take un/expected turns. Through local histories, rhetorical inquiries can engage submerged perspectives and open up "other sides" existing within modernity/coloniality. In adopting colonial difference from the borderlands work of Gloria Anzaldúa, Walter Mignolo continually emphasizes the importance of location as an invitation into border thinking that is so characteristic of San Antonio's politics. As Mignolo notes, the basic condition of border thinking is "the moment you realize (and accept) that your life is a life in the border, and you realize that you do not want to 'become modern' because modernity hides behind the splendors of happiness, the constant logic of coloniality."[32] As the case of COPS shows, praxis from colonial difference and de-linking from the assumptions of modern/Western epistemologies does not necessarily mean a complete rejection of modern/colonial systems but rather a de-privileging of Western Eurocentric systems of thinking and doing. This is why decoloniality is fundamentally an "option."[33]

In defining rhetoric as situated public discourse, Wanzer-Serrano notes the unique values of rhetorical approach to de/coloniality are its appreciation for situatedness, which demands, at a certain level, "the abandonment of abstract universals and ahistorical theorizing in preference for grounded theorizing that is attentive to the spatiotemporal and . . . embodied emergence of particular discourses in a 'multiplicity of overlapping contexts.'"[34] In attending to the grounded theorizing of divergence in diplomacy, the relationship among colonial/subaltern knowledges is only ever characterized by diffraction where the subaltern can refuse assimilation but also does not become only a governed subject, or a citizen gaining sovereignty through consumption. Divergence attends to the multiple possibilities of becoming otherwise within state colonial control, and diplomacy is the sophistic, precarious, and unending differential movement through these distinct possibilities. In their most recent work, both de la Cadena and Stengers clarify that divergence is not cultural difference but the constitution of practices as they emerge in their specificity and with other entities and practices.[35] Stengers emphasizes the way practices have their own world-making, and while conflicts of interest are the general rule, the remarkable events "are the creation of symbiosis or the weaving of coevolutions—that is, the making of connections between 'beings' whose interests, whose ways of having their world matter, diverge but who may come to refer to each other, or need each other, each for their own 'reasons.'"[36] In short, divergence is a trope for an ecological sense of diplomacy that is fragile, partial, full of conflict, and yet interdependent.

The work of Gloria Anzaldúa and AnaLouise Keating in *This Bridge We Call Home* is invested in similar concepts of interdependence across human and human/nonhuman communities. For Anzaldúa, to bridge is the work of a nepantlera to "attempt community, and for that we must risk being open to personal, political, and spiritual intimacy, to risk being wounded."[37] These interdependencies align with the work of queer Chicana feminists who draw from Jane Bennett's vital materialisms to include nonhuman relations in "broad considerations of coalition and justice . . . [and] a consideration of agency as an always contingent and contextualized relational practice as well as a possibility for action."[38] In this light, it is striking to compare Stengers's definition of cosmos and María Lugones's expansive definition of coalition—"always the horizon that rearranges both our possibilities and the conditions of those possibilities."[39] Coalitions are cosmic creations—in Lugones's emphasis on "the horizon that rearranges" and Stengers's emphasis on a capacity for articulations, one hears the creativity and excess of composing coalitions. In Karma Chávez's words, coalitional politics is "activism on the edge of realism," and its horizons are not utopian but rather "a present vision and practice that is oriented toward others and a shared commitment to social and political change."[40] Much like Stengers's emphasis on "events" in the composition of a cosmos, these orientations and shared commitments are conditioned by what Chávez calls "coalitional moments"—the spatial and temporal "turning points" or "junctures" that imply a coming together or connection whereby there is a possibility for change.[41] In the language of coalitions—horizons, junctures, and turning points—one hears metaphors of arrangement whose excesses mark emergence and divergence through pluriversal praxis.

Thus, divergence and diplomacy are viable theoretical concepts for thinking through the conditions of coalitional politics at level of praxis. The shared emphasis on situated praxis, the marking of divergences within ontological pluralism, and the goal of pluriversality are among the many reasons why Mignolo continues to point to the work of Latour, Stengers, and other nonmodern thinkers to note that "decolonial thinking is akin to nonmodern ways of thinking grounded on cosmologies of *complementary dualities* (and/and) rather than on *dichotomies* or *contradictory dualities* (either/or)."[42] This joint kinship is possible because quite independently nonmodern ways of thinking have long recognized modernity as a regionalism becoming a universal that elides alternative modes of living to the destructive extractivism and oppressive domination on which coloniality is based. Meanwhile, decolonial scholars continue to bring subalternized perspectives to bear on

the darker side of the politics of modernity/coloniality. In the relation between the nonmodern/decolonial, "conflicts of interest are the general rule,
but the remarkable events (without which only the triviality of predator-prey
relations would exist) are the creations of symbiosis or the weaving of coevolutions . . . [whose interests] . . . diverge but who may come to refer to each
other, or need each other, each for their own 'reasons.'"[43] The nonmodern/
decolonial are perhaps best characterized by affiliation not filiation, which
is cause enough for their divergences, and also an intellectual hybridity of
complementary dualities at in/appropriate(d) sites and for in/appropriate(d)
moments.[44] Divergence and diplomacy are thus productive tools for a pluriversal rhetorical praxis of coalitional politics.

To ground these concepts in situated contexts, what follows is a pluriversal rhetorical analysis of one of the most significant political achievements in
the history of San Antonio. Not coincidentally, this political achievement is
also partially the result of coalitional politics, a crowning legacy of both a local environmental group, and the Chicana/o movement. Rather than understand this achievement through critical cultural studies or social movement
theory alone, as other scholars have done, I look to this history to examine
coalitional politics as configurations of materiality and meaning through the
pluriversal concepts of divergence and diplomacy as a joint kinship among
nonmodern/decolonial approaches to coalitional politics.

EXCESSES OF RADICAL RHETORICAL PRAGMATISM:
COMMUNITIES ORGANIZED FOR PUBLIC SERVICES

By the mid-1970s, San Antonio's politics were dominated by what was known
locally as the "Black Hand" over San Antonio—a symbol of racism within
the power bloc of organized money via local bankers, lawyers, and old society families that deliberately kept the growing Mexicano/a and Latina/o
population from coming to power in San Antonio.[45] By most every account,
Texas was slow to acclimate to civil rights legislation, even though the landmark Voting Rights Act was signed into law by the Central Texan president,
Lyndon B. Johnson. State and city voting laws were particularly egregious.
Even by 1975 no printed voter information was available in Spanish, which
had the same effect as a southern literacy test.[46] While military installations
in the 1930s and 1950s provided San Antonio with a small Mexican American middle class, even by the 1970s people of Mexican origins were politically disenfranchised at the city level—so much so that leading Latina politicians and activists famously claimed there were seventeen white men who
ran the city.[47] In the context of national civil rights movements, it was dramatically apparent that the city's political representation did not reflect the
social, economic, and cultural environment of its people. Thus, the grounds

for this representation to be questioned, challenged, and transformed carried new potentials.

The story of the transformation of San Antonio's politics cannot be dissociated from the history of the city's Chicana/o movement, which began through the activism of social workers and college students working with Westside barrio youth. As David Montejano explains, early Chicana/o organizations like the Mexican American Youth Organization (MAYO), who were inspired by Black liberation movements, and particularly the Black Panther Party, interrupted a cycle of barrio warfare by identifying with an overarching race-ethnic Chicana/o identity and Chicana/o nationalism. As this movement evolved, it gave rise to group-specific organizations, specifically those that focused on electoral politics—such as the Raza Unida Party (RUP) and Mujeres por la Raza Unida—and those that focused on community sovereignty such as the Brown Berets and Chicana Berets.[48] By the mid-1970s, the Chicano movement in Texas faced internal divisions about leadership and tactics (especially as these topics were refracted through race, class, and gender) and external pressure from police and politicians alike. Yet these Chicana/o organizations also had several accomplishments, including the opening up of universities to Chicana/o students and the establishment of Chicana/o studies programs nationwide.[49] As Montejano writes, these organizations laid the groundwork for a second-generation of Chicana/o movement organizations that while they "deemphasized or distanced themselves from militant Chicano rhetoric, their fundamental objective remained securing equality and justice for marginalized communities."[50]

In this political context a young man recently trained in Saul Alinsky's Industrial Areas Foundation (IAF) politics, Ernesto Cortés, moved back to his hometown of San Antonio inspired by Cesar Chavez and the broader civil rights movement. He began his grassroots community organization with a two-year-long, door-to-door listening campaign through the Westside and Southside of San Antonio. Understanding that strong community organizations last only when the issues are defined by the locals and anchored by institutions with roots in local communities—namely, the Catholic parish networks of San Antonio—Cortés sought to identify and train community leaders willing to exercise their civil rights through local democratic participation. And the primary issue that these residents of San Antonio had in the mid-1970s was the interrelated issue of flooded neighborhoods, drainage, and infrastructure investment in the urban core. Having finished his listening campaign, and with modest start-up funds, Cortés invited community leaders together to organize what was first named the Committee for Mexican American Action, but was later renamed Communities Organized for Public Services, or COPS.[51] The name change is significant—in discourse it strategically downplays social action through Mexican American identities, yet in practice it links these identities to an organization "for public services"—for investment in flood protection, in neighborhood improvements, and clean water—all the while playfully identifying, through a *détournement*, as the true civil servants and protectors of the people, the COPS![52]

Through COPS, Cortés effectively brought Saul Alinsky's version of radical pragmatism, with its foundations in sophistic rhetoric, to the Southside and Westside neighborhoods of San Antonio, Texas.[53] Chávez makes a point that Alinsky's radical pragmatism could reinforce problematic norms for coalitional politics, but she also notes that these do not exhaust Alinsky's potentialities for radical diplomacy. His emphasis on training leaders for flexible personalities, quests through uncertainties, and political relativity is highly associated with a sophistic diplomacy as "the precarious and unending work of traveling to and from multiply nested but nevertheless distinct realities."[54] Like sophistic rhetoric, then, diplomacy starts from where the world is rather than where one would like it to be, but rather than just a world of human communities, nonmodern diplomacy understands material conditions as causes for thinking about the emergence of diverging minorities that can make political decisions as difficult as possible. As Robert Danisch has noted, pragmatism reiterates many of the intellectual commitments of sophistic rhetoric in new variations,[55] but from the perspective of a nonmodern/decolonial diplomacy, contemporary examples like COPS necessarily exceed the boundaries of classical rhetorical theory in such a way that the excess provides a useful resource for rethinking rhetoric in community organizing and coalitional politics.

For a pluriversal rhetorical praxis, this formation poses at least three questions: First, how can the excess of materialities as causes for political thinking be applied to rethink coalitional politics? Second, how do the valences of Chicana/o publics exceed, resist, and/or challenge rhetorical concepts to account for thinking at the exteriorities of coloniality? And third, how can de la Cadena and Blaser's notion of the uncommons capture how the excesses of heterogenous worlds and practices come together though they only strive to be what they are? In the next section, I take up each of these questions through the language of divergence in diplomacy as pluriversal rhetorical praxis for coalitional politics.

DIVERGENCE AND DIPLOMACY FOR THE EXCESSES AND UNCOMMONS OF COALITIONAL POLITICS

By organizing the Westside and Southside of San Antonio through the Catholic parish networks, Cortés marshaled the lived experiences of working- and middle-class Mexican Americans into an organized political threat by bringing subalternized discourses and practices into articulation with dominant city logics.[56] But the collective becoming of the politics of COPS was not an achievement produced by humans themselves. With the help of Ernesto Cortés and others, neighborhoods and parishes began to define their own issues of concern, and the primary issue on people's mind was flooding. Because they thought from where they stood,[57] and because the situation of generational flooding generated the power to make them think,[58] these newly organized neighborhoods became coalitional publics by turning flooding events into causes for political thinking about the city's inequitable arrangement of public services. Such political emergences are not isolated from the ecology in which they arise; rather, it is the specific relations to urban waterways that participate violently in peoples' lives that give material force to the necessity of coalitional politics.

So, when the flash floods arrived, this time they were not only figured as racialized oppression, in addition, they were made meaningful through the potentials of Alinsky's radical power politics that were actualized through the constraints and obligations of a multigenerational experience of flash flooding. On May 7, 1974, a heavy rainstorm hit San Antonio causing 150 Westside residents to flee for the night. Three months later, COPS organized their first public meeting with the city manager. Armed with Alinsky's methods of mass power politics, COPS showed up to the meeting at a Westside high school with five hundred members in attendance. In the COPS press releases sent out to motivate this mass bloc of people for a meeting with the city manager, Ray Kaiser wrote about the relationships between materiality, precarity, and the city's development priorities:

> The only reason that the flooding continues that is responsible for the loss of life and extensive property damage every time it rains is because the city has neglected the Bond Issue Drainage Projects that were funded in the 1970 Capital Improvement Program. . . . We think that property damage to home owners, and the prevention of tragic deaths from killer floods are more important than Repairs on the Convention Center, [sic] We recognize that many of the North Side projects are worth while, but we think that the developers can wait their turn, as we have had to wait, until after these crucial projects are finished. . . . There is no excuse for a city as prosperous as San Antonio to have the kind of flooding problems that exist in the southside and westside of San Antonio. . . . We will not rest until our streets are passable and our homes are safe from these destructive waters.[59]

Kaiser's exhortation about killer floods can be read in a number of ways, but through divergence and diplomacy, it is thinking with flash flooding as it emerges in its specificity and yet with other entities and practices that allows

Kaiser to move through multiply nested but distinct realities among developers, residents, and bureaucrats. In attending to the divergences of flooding events—the partial connections that mark colonial difference—like loss of life, property damage, and city neglect, Kaiser invents a public standing for people living in close relationship to killer floods. From there, he engages in diplomacy as a practice of border-thinking: "the developers can wait their turn, as we have had to wait"; "many of the North Side projects are worth while, but we think"; "the city has neglected the Bond Issue Drainage Projects." Through divergence and diplomacy, COPS becomes what they are not without becoming what they are not. In practice, this meant that five hundred COPS members became a political force at various city meetings on flood control without becoming traditional politicians. Kaiser's focus on the arrangement of materiality and meaning rearticulates subaltern bodies, discourses, and material environments through the coalitional politics of organizing a massive bloc of people that forces the city manager to think in the presence of those most affected by inequitable material conditions.

This meeting and those that followed have taken on a legendary status among those who know the history of city politics in San Antonio, partially because these meetings were the first of what still exists today as the COPS accountability sessions. At the meeting Ray Kaiser helped organize with the city manager, five hundred COPS members demanded *yes* or *no* responses, and tangible action on drainage improvements with little patience for quibbling and excuses. One of the early leaders of COPS showed slides depicting the havoc that torrential rainstorms had caused in the neighborhoods and told the city manager: "Scenes like these have been there for years and are still the same. We have decided not to take it anymore. We have decided to make our problem, your problem."[60] At the city council session the following week, COPS again showed up with hundreds of members who disrupted the typical decorum by refusing to sign in one at a time; instead, they demanded action be taken to protect their neighborhoods. Their main spokesperson was Mrs. Hector Aleman. As she took the podium, every COPS member stood in support as she told the story of how every time it rained even a half inch, residents had to shovel out water from their living rooms. Houses, streets, local parks, and neighborhood churches all were underwater. "How would you feel getting out of bed in the morning and stepping into a river right in your house?" Mrs. Aleman exclaimed as her supporters roared.[61] Then she pointed out that the infrastructure project that would have protected her neighborhood had been designated for funding back in the city's master plan of 1945, yet it had never received one cent. When the newly elected mayor was told that the flooding in these neighborhoods affected upwards of 40,000 people, and when COPS threatened to mobilize a massive voting bloc, he gave the city staff four hours to find the funding for the infrastructure project. Over the next few weeks, COPS relentlessly kept pressure on city officials for more investment. In the subsequent bond election cycle, their neighborhoods received $46 million to fund fifteen Westside drainage projects.[62]

In these early meetings with city officials one witnesses the first excess of the radical rhetorical pragmatism of COPS—not more deliberation, but rather an embodied and coalitional power politics through organizing a massive number of disenfranchised people around floods as causes for thinking about politics. *In generating the power to think with the flood, COPS became the flood.* The fleshy subjectivity of a massive number of bodies at a city manager meeting, along with the research that reminds the city manager of his obligations to neglected communities, compels a different form of diplomacy by thinking from the divergent experiences of the subaltern.[63] By bringing subalternized discourses and practices into articulation with dominant city logics, COPS used city meetings as sites of partial connection that are also sites of divergence, especially by holding local politicians accountable while also never becoming, or even endorsing, politicians. Through this fleshy subjectivity COPS engages in diplomacy as a matter of bordering and calling into question the wisdom of a given arrangement of public services like flood protection. Through diplomacy as border-thinking, COPS reshaped the very formulation of the problem as a lived flux with no demarcated borders, but where floods become a cause for political thinking enacted at opportune times that allows an alternative form of rhetoric and politics to emerge. Through divergence and diplomacy, one witnesses how coalitional politics are radically situated through obligations and constraints of a shared *oikos* that helps give shape to a collective ethos—a coalition as the horizon that rearranges both our possibilities and the conditions of those possibilities.[64]

The second excess that a pluriversal rhetorical praxis will account for is located in the high correlation between poverty, race, gender, and location in San Antonio where the hardships of inequalities fell on Mexican American women the most. Yet, in grounding radical rhetorical training in the embodied locations of neighborhoods and parishes, COPS effectively transformed Mexican American women into political organizers and leaders: "From the beginning, women did most of the research. They set up the meetings, made the telephone calls, and got out the vote. . . . It was a new role for Hispanic women, who, with few exceptions, had never become active politically or in the women's movement."[65] COPS leader Toni Hernandez made these connections among proximate material conditions and power politics explicit: "In the Mexican culture, women have the responsibility for home and children. Schools related to children. Neighborhood is a part of home. So, if a woman raises hell about drainage in the neighborhood, that's okay."[66] Beatrice Gallegos, a president of COPS from 1976 to 1978, wrote of what it meant to the women who participated: "Not only has the organization changed the city of San Antonio, it has changed us as individuals. It has developed us to make us better persons in our own families, in our churches and in our communities."[67] In short, what COPS succeeded in doing was offering Mexican American women a chance to develop their public leadership roles by connecting Mexican American cultural values to the local politics of neighborhoods and communities. In this personal transformation, home is not just a private domain but a public practice of heterogeneously entangled worlds; children are not just individuals, but a public entity entangled through systems of education. These are the pathways of transformation for the city (an *oikos*) and its inhabitants (an ethos).

In light of divergence and diplomacy, it is the articulation among bodies, location, and practice that allowed this new politics to emerge—care for homes, schools, neighborhoods, and children were now entangled with a politics of inequitable public services in ways that made it okay for Mexican American women to raise hell. As a performative articulation of adaptation without assimilation, of divergence within diplomacy, these women now move among multiply nested realities not to negotiate different perspectives but to reconsider relations across distinct worlds from the embodied locations of the colonial periphery. Yet these women are not simply sophists. By leveraging power politics through the arrangements of public services, what also matters is a collective becoming that humans could not produce "'by themselves' but only because of the situation that generated the power to make them think."[68] Through divergence and diplomacy, Mexican American women think with embodied locations to radically restructure the boundaries of belonging to publics by becoming self-different— they become other than what they were, while continuing to be the same.[69] These coalitional politics are not just acts of collective will but of the lived flux of moments and events that compel the conditions for community organizing.

These coalitional politics effectively blamed the experience of destructive flash floods on greedy developers and bought-out bureaucrats, and they went a long way toward establishing COPS as a viable political force in citywide politics. Once the investments in infrastructure were made, the attention of COPS shifted to other areas of equity in public services, particularly higher utility rates for city residents in order to subsidize infrastructure for developer-driven suburbanization. These new fights over utility rate increases for its neighborhoods meant COPS became directly involved with

San Antonio's water politics. These politics demanded a different emergence of coalitions. In October 1975, when the city council voted to permit construction of a large regional shopping mall on 129 acres over the Edwards Aquifer recharge zone, COPS channeled their outrage by linking two somewhat disparate discourses and practices—protecting the purity of the aquifer water and ending the neglect of the urban core—both of which were threatened by suburban development. By the time of the annual COPS convention that November, COPS had fused practices of core urban development and protecting water quality in order to galvanize members of COPS against a common enemy. Press releases, flyers, and letters linked these arguments and urged members to act: "The question is who will benefit and who will pay for these ridiculous policies that force us to pay higher water bills, gas and electric bills, sewerage charges—to subsidize development that will: 1. Poison our water 2. Destroy our neighborhoods 3. Turn downtown into one big parking lot, [. . . the bankers and developers] will benefit while we pay. The COPS Convention is a <u>MUST</u>. The survival of our neighborhoods and our city is at stake."[70] The development of another regional shopping mall over prime water recharge areas provided COPS with an event—a change in *oikos* and therefore a change in ethos—that offered a coalitional moment for COPS to take its tactics outside its neighborhoods and into broader city politics. In linking the destruction of urban neighborhoods to poisoned water, COPS created a site for what were at the time disparate political practices.

Thus, at the end of their 1975 convention, and in response to the proposed development of the shopping mall, COPS formed a coalition with the Aquifer Protection Association (APA), an environmental advocacy group that arose out of the League of Women Voters.[71] In jointly identifying developers as common enemies, the coalition launched a successful petition campaign that placed a referendum on the council's decision to allow for the mall's development. The significance of this coalition is found not just in the grassroots community organizing through petitions and voter turnout drives; the larger significance is found in how COPS combined the practices of caring for and protecting urban neighborhoods with caring for and protecting clean water as a method for making suburban development as difficult as possible. The event of suburban development actualizes the obligations and constraints of coalitional politics in ways that cannot be completely predicted. Although COPS and APA had formed a coalition to stop the development of a shopping mall, it's clear that both organizations had different reasons for doing so. While in the COPS archives one can clearly see a rhetorical kinship between pure water and urban core investment, as in the Sarabia quote above, only the former is emphasized in the APA archives. Thus, this coalition was characterized by interests in common, which are not the same interests, or what de la Cadena and Blaser call an *uncommons*—a negotiated coming together of heterogenous worlds and practices that strive to be what they are, which is also not without others.[72]

Why a local environmental organization like the Aquifer Protection Association refused to engage in the discourses and practices of urban core development as a method for stopping suburban expansion is also a question of why environmental organizations refuse to be with others in efforts of environmental protection. This lack of reciprocity, and aversiveness to the urban core, are forms of undeserved privileges that Wanzer-Serrano points to in the epigraph to this chapter. In refusing to engage these arguments, and in refusing diplomacy as matters of arrangement also for the urban core, the APA problematically refused to create a world of many worlds in relation to clean water. It speaks to COPS that this aversiveness did not impede coalitional politics, and yet it also possibly speaks to this coalitions' lack of endurance.[73] It is here at the height of this coalition that I will end this narrative on divergence and diplomacy to capture the excesses and uncommons of coalitional politics. But before I conclude, it's worth summarizing the effects of this coalitional practice in light of the political forces working outside of local conditions. After summarizing these effects, I then conclude by reflecting on the implications of divergence and diplomacy in coalitional politics for a new era of climate politics.

COALITIONAL POLITICS AND FEDERAL LEGISLATION FOR LOCAL TRANSFORMATION

Through the coalitional politics of COPS and the APA, San Antonio voters approved a referendum to stop the shopping mall development on environmentally sensitive land. But while local politics proceeded through scientific studies and the city-state legal system, on the federal front, in the fall of 1975, Texas was brought under the jurisdiction of the 1965 Voting Rights Act, which meant that San Antonio's annexation of surrounding territory was scrutinized for its impact on voter representation in city government. Based on a lawsuit filed by another second-wave Chicana/o organization— the Mexican American Legal Defense and Education Fund (MALDEF)—the justice department required the city to rearrange its election system, and a 10–1 plan, where the city was divided into ten districts with a mayor chosen at large, was offered to city voters in the next election cycle in January 1977.[74] Thus, while the water quality issues awaited solutions from technical study and/or legal decisions on the city council's referendum, the politics of voting rights provided a previously unforeseen avenue for exercising legal response-abilities around urban core investments and clean water.

In January 1977, in an extremely close vote, the 10–1 city council plan passed 51 to 49 percent. COPS and APA not only delivered blocs of voters, but also the political coalition and the entire aquifer debate effectively persuaded Anglo middle-class voters that developers were not in their interest and many stayed home. The result was that in April 1977, the new district structure yielded a very different city council—five Mexican Americans, one African American, and an Anglo female from the Southside. All told, "seven persons were elected from areas within the city which had experienced little or no representation during the previous two decades."[75] The coalitional politics for clean water and urban core development clarified the issue of city representation for many voters, who because of this local fight increasingly saw the connections among water pollution and inequitable development as an issue of controlling developers and channeling city suburbanization. In short, the local fight for water security and the national fight for voting equity provided a pathway necessary to bypass state resistance to city governance.

While the Voting Rights Act can be seen as imposing change on city politics from the outside, the actual vote on the plan was the culmination of a series of local practices: an extensive legal campaign by MALDEF, an environmental controversy that brought developer and technocrat collusion to light, and a realization of political power by long-frustrated Mexican American voters. As David Montejano and Peter Skerry note, the Voting Rights Action could not "*will* Mexican-American power into being."[76] The political will necessary to establish a 10–1 city council structure was catalyzed through the coalitional work of COPS and APA. For environmentalists, the effect of this new city council structure was a temporary ban on development over the aquifer. For COPS, the effects were the ability to exert influence over five districts where COPS had a presence, and through the Community Development Block Grant process, they could now channel federal dollars to urban core development. Over the next decade, COPS secured more than $500 million for Westside and Southside improvements, and one of San Antonio's own, Henry Cisneros, would be elected as the first Latino mayor of a major metropolitan city.[77] In short, the coalitional politics for clean water and equitable public services helped transform San Antonio by helping to create a city council that reflected a coalitional form of politics established by COPS and the APA.

What does this fairly traditional story of democratic republic governmental mechanisms mean for the broader narrative of divergence and diplomacy in pluriversal rhetorical praxis? Throughout, I have defined this praxis as the situated public discourses and practices that compose ecologies of practice across heterogeneously entangled worlds as pathways toward political transitions, transformations, and/or intransigence. Through divergence and diplomacy, I have attempted to demonstrate how this story of political transformation was not just an achievement of community organizations, Vatican II, or cultural heroes. Divergence and diplomacy captures the necessary excess of worlds entangled with the human—the landscapes, watersheds, airsheds, animals, and more. The political coalition among COPS and the APA could not have arisen anywhere but through the pluriverse of San Antonio. And the larger goal of this coalition was to refract the course of one-world development into many worlds that still negotiated their difficult being together in heterogeneity.[78] Divergence and diplomacy are un/common places with their un/expected turns, yet still capable of being traced as an ecology of practices across entangled worlds. COPS and the APA did not engage in coalitional politics because they had the same facts, or even shared the same concerns, but because they witnessed human/more-than-human environments that needed care and protection against violent one-world development modalities. When these kinds of fleshy responses are paired with political savvy and with legislation that itself is a federal enactment of creating and caring for multiple worlds, the potential for a political transformation that fosters pluriversal worldings becomes newly possible.[79]

PLURIVERSAL HISTORIES FOR A NEW ERA OF CLIMATE POLITICS

Yet, of course, the achievements of coalitional politics and the Voting Rights Act were soon eroded by the ability of the courts to legally protect one-world amalgamation and capital. The local landowners who stood to make money from the development sued the city for $1.5 billion in damages. By January, a Texas court ruled the ban on the supermall development over the aquifer was illegal because the state constitution omits referenda as tools of citizen action. Soon the total ban was stripped down to a limited city ordinance that required a detailed natural resource disclosure statement to show how their project would not negatively affect the aquifer. Anyone who didn't comply wouldn't get water. So suburban growth certainly did not stop, but rather, development could now be lightly channeled through city-level regulations that remain largely in effect to this day.[80] In the end, the story of COPS/APA is another enduring legacy of climatic response, coalitional politics, and pluriversal praxis in San Antonio, even if the successes of these responses were attenuated by legally protected capital.

What do these histories of rhetorical praxis mean in a new era of climate breakdown that will be pluriversal in its effects? What can these local histories of human and more-than-human response tell us when the neighborhood becomes the locus of an adaptive and resilient response to unprecedented shifts in prevailing climatic and political embodiments? I believe partial answers to these questions can be captured by divergence and diplomacy in coalitional politics. Given the speed, severity, and scope of the problem, there is likely to be a massive mobilization of people fighting to protect their livelihoods by striving to be the best of what they are, which is never without human and more-than-human others. One can already witness this emergence in the global climate strikes of September 2019. Coalitional responses will likely take on divergence and diplomacy at sites of public services—water, energy, transportation, homes, and more—as matters of arrangement, as a differential movement among multiply situated worlds to reconsider their relations to each other and to public services. Turning public services into causes for political thinking also means climate politics will arise from perhaps unexpected publics—the family, the public school, the neighborhood, the community organizations, the church, the working class. Climate politics will necessarily become everyone's politics—a pluriversal politics living through climate breakdown and fighting against one-world dominance of fossil fuel extractivism.

Therefore, maintaining a sense of the political in public services matters. COPS was not some weekend walk event for a good cause. It was the radical and pragmatic organization of a massive number of working-class people to take political power by becoming the flood that could hold its political representatives accountable. Today, COPS has lost much of its power politics. Instead, major developers and businesses hold city politics hostage by threatening to take their investment money elsewhere if "mob politics" make San Antonio's political climate too unstable. Facing this dilemma, COPS and its supporters sacrificed mass power politics in favor of unequal capital development that still provided a barely rising standard of living through minimal job creation. But while its radical power politics are dormant, COPS is still an organization that puts into practice the notion that serving the poor also means protecting the environment, as the epigraph from Pope Francis's encyclical *Laudato Si'* articulates, and as seen in the COPS murals in figure 5.

For contemporary climate politics in San Antonio and beyond, this case study demonstrates that coalitions of climate mayors must themselves engage in practices of divergence and diplomacy that center community organizations in the policies around "climate proofing" the neighborhood. Metaorganizations like COPS have much to teach climate adaptation leaders about building neighborhoods with strong social networks, and revitalizing civic and religious institutions, partially through integrating early warning systems and planning for systems of aid and support. While it is unlikely that transformative climate politics will occur without some federal legislation, local politics are just as necessary for transitional and transformative futures of climate politics. Consider: if the coalition of over four hundred

Figure 5. COPS murals at Cassiano Homes. Photographs by Kenneth Walker.

climate mayors can simply honor existing commitments and policies at the state and local level, it will get the United States two-thirds of the way to the goals of the Paris Agreement. At the same time, it is the pluralism inherent in notions of adaptation and resilience that are characterized by an ecology of practices that contain the capacity for transformation and unexpected turns since one can never know what practices will partially connect and what new politics will emerge. The wily unruliness of divergences in partial connections and the potential articulations of which they are capable are reasons to keep practicing rhetoric.

Place-Keeping and Rhetorical Inhabitation as Pluriversal Adaptation

We have generational poverty for decades on the western side and the east side, and so, you know, dealing with these issues, it's like "but we want to put up a hotel and a park and say hey San Antonio is great," and so I think it's an opportunity to push back against that narrative.
Gentrification: Standing with Soapworks and Towne Center, 12:15

As I write this chapter in the fall of 2019, the levels of carbon dioxide in the earth's atmosphere are higher than they have ever been in the last 800,000 years. This means the very climatic conditions that gave rise to human evolution, let alone human civilization, are now significantly altered.[1] This unparalleled planetary experiment will have consequences that cannot be completely foreseen, and potentially, climatic history will not serve as a reliable guide for the climatic futures to come. In a best-case scenario, future climates will resemble the extreme weather events of the past occurring more intensely and more frequently—the one-in-one-hundred-year event that now happens every few years. In a worse-case scenario, this experiment will create a deadly and unrivaled experience of supercharged weather that leaves cities completely unprepared for the future—the one-in-five-hundred-year event that now happens every year, or worse, the multiple catastrophic weather events that hit simultaneously every few years. Regardless of the exact scenario, it is certain that these events will put tremendous pressure on local governments already stretching to provide basic public services to their inhabitants. So, beyond the daunting task to mitigate carbon pollution and transition to clean energy, cities are now facing the realities of adapting to the deep uncertainties of multiple possible future worlds under climate breakdown.[2]

Like many cities behind the climate curve, San Antonio is just now considering preparations for climate change through a citywide climate action and adaptation plan (CAAP) with two major components: further mitigation of the city's CO_2 emissions, and an adaptation plan to prepare for current and future extreme weather. Scholarship situates the contemporary emphasis on climate adaptation within the context of the overarching failures of sustainability politics.[3] Where sustainability was meant to be a progressive move toward stopping the consumption and waste of natural resources in order to maintain ecological balance, adaptation is a more defensive posture that accepts the current and future consequences of human-induced climate change (while simultaneously working toward mitigation). But like sustainability, the main problem with climate adaptation is that it is easily adopted into the status quo, and it often ignores the necessity of confronting the power dynamics behind leaving carbon in the ground, organizing against oil and gas pipelines, or shutting down coal plants. In response, activists and scholars have linked adaptation to social reform and forwarded a number of concepts to keep climate adaptation appropriately political—deep adaptation,[4] just transitions, and transformative climate adaptation.[5] It is time to ask how projects for climate adaptation can help expand public spaces and nurture a sense of belonging in ways that address the legacies of coloniality and anticipate the extreme weather events to come.

Of course, adaptation projects are never without nonhuman others, and they are never outside the contexts of global development. Truly deep, transitional, and/or transformative climate adaptation projects must also reckon with its de/colonial histories, its contemporary politics, and foster pluriversal visions of belonging to a place, a home, a world among many worlds. The last two chapters unearthed rhetorical histories to highlight the importance of shared sacrifice and coalitional politics in pluriversal rhetorics responding to extreme weather events. I have argued that while the climatic past of San Antonio may not have bearing on its future, the histories of climate politics as a response to extreme weather can certainly inform current and future practices. And for river cities like San Antonio, contemporary adaptation often takes the form of urban watershed restoration projects, which anticipate extreme floods and reimagine urban core development. In this sense, adaptation projects bring together ecological and development rhetorics by thinking with climatic futures to reimagine a place-based sense of belonging.[6] While some excellent rhetorical work is currently being done at the intersections of ecological restoration and de/coloniality, it is clear that rhetorical scholars need more tools to keep these projects appropriately political so that they also do the work of reconstructing pluriversal relations among worlds through climate crises. In other words, this chapter asks the question, how can pluriversal rhetorics reckon with contemporary forms of modernity/coloniality in local adaptation and ecological restoration projects?

To highlight the politics of climate adaptation, ecological restoration, and their pluriversal potentials, this chapter engages in a rhetorical analysis of a contemporary urban watershed redevelopment and restoration project in downtown San Antonio—the San Pedro Creek Project (SPCP). The project is sited along a two-mile stretch of a creek on the Westside of downtown, and its urban-center headwaters were the site of the first settler colonial contact in this valley in 1718. Through the SPCP redevelopment project, city and county officials plan to transform the creek from an urban drainage ditch into a world-class linear cultural park. As a creek restoration project, climate politics are constantly foregrounded through the specter of flooding as a cause for political thinking about cultural, ecological, and economic diversity. Yet, as local governmental institutions have become more representative of San Antonio's dominant Latinx public, the exact problem becomes how to build, design, and maintain a pluriversal vision of contemporary San Antonio lifeways as the necessary conditions for addressing the historical and contemporary effects of one-world colonialism. I engage this site through the un/commonplaces and the un/expected turns of place-keeping and inhabitation in order to show what is at stake in these projects—a renewed sense of adaptation as a cosmic transformation from urban disenchantment toward urban reenchantment, from displacement toward replacement, and from economic segregation toward ecological and economic integration that is only possible when climate adaptation is kept appropriately political through pluriversal praxes.

To do this, first I situate the San Pedro Creek Project within the contexts of development rhetorics, pluriversal design, and the global demands for cities to attract consumer markets for global capital investments. Then, through a rhetorical analysis of the design and implementation of the first section of the park, I examine the un/common places of authenticity and argue for place-keeping as a pluriversal rhetorical praxis that resists the forces of public participation and restoration ecology when they reify white spatial imaginaries.[7] Then, I examine the rhetorical work of a housing justice advocacy group that somewhat successfully resisted gentrification along the project through discourses and practices of inhabitation as re-placement rather than displacement. Together, the topologies and tropologies of inhabitation and place-keeping help maintain pluriversal designs through transitional climate adaptation projects. These climate politics reveal that the San Pedro Creek Project is more than a new River Walk for locals. Rather, it is a project that uses its deep de/colonial, ecological, and climatic histories to reimagine

development with others as the necessary conditions for reconsidering relations among worlds in crisis in the spaces we live, work, play, and worship.

DEVELOPMENT RHETORICS AND PLURIVERSAL
DESIGNS IN THE GLOBAL GROWTH MACHINE

A city's relationship with its waterways reflects its histories and current visions for itself, especially when contemporary urban redevelopment is centered on watershed restoration. But, of course, redevelopment trends are
never just about local histories and identities. Urban watershed restoration,
also called daylighting, or the "uncovering of streams buried under layers
of pavement . . . [as] a way of creating green corridors and cutting down on
pollution in cities," is a global phenomenon.[8] Certainly, daylighting offers
many local benefits—improved urban hydrology, urban corridor habitat, bio-
diversity, and cultural and aesthetic amenities. But quite clearly, local governments that invest in restoration projects are doing so with the hope of
attracting consumer markets for global capital investment. Thus, ecological
restoration cannot be extracted from the movement to place local ecologies
and local cultural expressions at the center of global economic development
strategies. Indeed, as community scholar Marisol Cortez has written, San
Antonio's push to develop its downtown urban corridors are a manifestation of what John Logan and Harvey Molotch call global growth machines,
"a coalition of business and civic interests that promote economic growth,
seemingly at any cost, as both a taken for granted good and proper object of
policy."[9] As I will do here for the San Pedro Creek Project, Cortez places former San Antonio mayor Julian Castro's "Decade of Downtown" initiative in
2010, and the designation of San Antonio's Spanish colonial missions as UN
World Heritage Sites in 2017, within the context of global growth at seemingly any cost.[10] And just as we saw in the last chapter, it is an open question
how effective cities can be in resisting and/or shaping the forces of global
development.

Rhetorical scholars like Jenny Rice, Lynda Walsh, Caroline Gottschalk
Druschke, Laurie Gries, and Candice Rai have long used topological methods for place-based, ecological, and development rhetorics. Such work argues that by focusing on sophistic traditions of topoi, the material-semiotic
circulations of commonplaces allow scholars to rethink and reimagine everyday public subjectivities, or "the way people are encouraged (through
exchanges of everyday talk) to imagine themselves in the public sphere."[11]
Tracing rhetorical topologies allows for an analysis of the creation of public subjectivities that has dramatic implications for procedural, normative,
and constitutive dimensions of democratic participation—examples range

from the circulation of the famous Obama Hope poster, to Indigenous imagery used for nation-building, to the images of ozone depletion that created analogical verifications of the body in pain.[12] Such technical and artistic images change the way everyday people imagine themselves in relation to the world and therefore reorient participation in public decision-making. Yet in pointing to the need to deepen materialist analyses of such rhetorical ecologies, Druschke points us to a trophic future of rhetoric where the symbolic, affective, and physical are treated as ontological coequals for the purpose of including other beings and things into rhetorical circulations.[13] I argued in the first chapter for the relevance of tropologies as un/expected turns to open up rhetoric's potential for new definitions that are ecologically emplaced through everyday interactions, always material and semiotic, and therefore of the world. My interest in politics leads me toward a broad definition of rhetoric as everyday public symbolic practice where presidential speeches are as rhetorical as plants turning toward the sun, even as their public performances are also characterized by divergence.[14] These un/common places and their un/expected turns connect rhetoric as discursive practice to planetary systems that always already involve novelty within repetition. When such tools are applied within pluriversal praxes, rhetoric grapples with the everyday coloniality of power. This view of rhetoric would seem to have a particular purchase on pluriversal notions of restoration in the design of place-based adaptation projects.

In its anticipation of more frequent extreme flash flooding, and in its situatedness on the west side of downtown San Antonio, the San Pedro Creek Project necessarily confronts climate politics through pluriversal worldmaking. Thus, the SPCP could be considered a case study of what Arturo Escobar calls designs for the pluriverse and transition discourses whose emphasis on place-making and collaborative practice are grounded in de/colonial visions *through* the ecological and social crises created by modernity/coloniality.[15] Focusing on the project's design as an aspect of transitional climate adaptation allows one to hold in tension the forces of the global growth machine with the local forces that animate public discourses and practices of place-making. In other words, when examining the SPCP project through pluriversal rhetorics, I engage a series of processes articulated by Escobar: "dominant modernity's negation of other worlds' difference; the resistance and excess constituted by subaltern subjects at the fractured locus of the colonial wound; and the challenges to the dominant modern core stemming from nondominant modern sources."[16]

But before I detail how place-keeping and inhabitation conduct this pluriversal rhetorical work, I want to emphasize a point I will return to throughout—while local ecologies and local cultural expressions are inextricable, they are also always conditioned by global economic developments. Viewed in these political economic terms, the SPCP seeks to transform a modern/colonial urban ditch into a "linear cultural park" in order to expand consumer markets, attract global investments, and accumulate land-based wealth, often under the facade of community empowerment. While such projects expand and enhance public space, that space is largely valued for the money to be made from it, not for the people who currently inhabit that space, especially if they have little money to spend. In this development model, growth is measured not in terms of sustaining healthy and vibrant communities already in place, but rather in terms of a city's ability to attract employers that grow consumer markets, often while displacing current residents. And yet, such political economic projects are never uncontested. If development projects and their rhetorics reflect a city's relations to its waterways, then what is it that animates community control? How can local publics push global development to invest in the people who inhabit these communities while also investing into creek restoration and city beautifying? Or will global demands for markets fully dictate the type of development and how quickly it happens? And how severely will these global markets continue a long legacy of colonial relations of structural inequality that create natural playgrounds for the leisure class while displacing working people and promoting paper-thin tourist versions of "cultural identity"? What follows in the next section addresses these questions through an analysis of the initial design of the SPCP as a practice of place-keeping.

LATINO URBANISM AND PLACE-KEEPING
FOR PLURIVERSAL ADAPTATION

The specter of flash flooding has always loomed over the SPCP, but not just as a funding source for building urban parks; from the beginning of the design process the creek was treated as a character in the unfolding de/colonial drama of San Antonio.[17] On a design level, the ways in which flooding was transformed into adaptive capacities for place-making are best situated within the architectural firm the county hired to design the construction of the SPCP—Muñoz and Company. One of the largest and oldest minority-owned design practices in the country, it is led by CEO Henry R. Muñoz, a San Antonio native who successfully launched a career in politics. Muñoz's record of entrepreneurship and philanthropy is impressive: he is the founding president of Texas Public Radio, former vice chairman of the Smithsonian, director of the Congressional Hispanic Caucus Institute, lead organizer behind the National Latino Museum on the National Mall, and finance chairman of the Democratic National Party. Muñoz and Company embrace a broad vision of architecture as "the exploration and expression of cultural identity to address the specific needs of historically underserved groups."[18] At its core, Muñoz's design is cultivated by a transhemispheric Latinx aesthetic that aims for "cultural synthesis"—an acknowledgment that "global connections generate ever-newer blended cultures . . . [with] new opportunities for original synthesis." In each project, Muñoz "champion[s] the recognition of convergent identities in community revitalization [and] embrace[s] this new creative environment with enthusiasm and our continuing dedication to design excellence."[19] Thus, in Muñoz's design philosophy one witnesses a borderlands theory of creativity through cultural hybridity, of the local as a synthesis of global forces, of democratic politics to meet the needs of the structurally marginalized, and an adaptive capacity for place-making through cultural design.

The cultural synthesis that Muñoz and Company sought in its design of the creek project is best described in their initial design plans, which drew inspiration from Latino Urbanism and San Antonio's cultural landscape to create five distinct creek character areas. The designers emphasized a need to forge a stronger bond between people and water and use the creek as a cause for reflection on the creek's rich history and its capacity to create a unique sense of place. The five distinct character areas of the creek were meant to function as "simultaneously segregating and aggregating . . . by drawing on the character of place, it is possible to celebrate these individual components and establish a sequence while integrating them into a whole."[20] This holistic integration is primarily achieved through Latino Urbanism—a Los Angeles design aesthetic based on the urban practices and living preferences of Latinx communities. For Muñoz, Latino Urbanism was about building community through the artistic languages of "a new America" that "blurs the lines of indoor and outdoor . . . synthesizes the common and the exquisite . . . [to] help us re-think, re-cycle, and re-use everyday materials [that] have been used in an architectural manner in communities with very modest means for generations."[21] This synthesis of recycling common, everyday materials with the exquisite draws inspiration from many neighborhoods on the Westside of San Antonio and beyond. Localizing Latino Urbanism to create an urban linear park filled with creek characters is also a statement about the possibilities of designing for the pluriverse through place-making in San Antonio.

But Muñoz and Company's initial designs for place-making were thoroughly contested when, through the local public participation process, they confronted what Roberto Bedoya via George Lipsitz has called a "white spatial imaginary."[22] According to this thesis, Latinx urban designers, and Latinx cultural expressions, continually confront the presumption that public place-making should be characterized through Anglo-American cultural aesthetics, which are largely assumed and invisible until they are confronted with difference. According to Bedoya, this "'white spatial imaginary' [fosters] an antiseptic ethos that effectively deem[s] being poor and of color as civic imperfections to be expunged."[23] In contrast, Latino Urbanism's emphasis on the reuse and rethinking with everyday objects from generations of people with modest means is an entire movement of Chicana/o art called rasquachismo. As Chicano art historian Tomás Ybarra-Frausto describes it, rasquachismo is a Chicana/o aesthetic with an "attitude rooted in resourcefulness and adaptability yet mindful of stance and style,"[24] which for Bedoya signifies an "imaginary structured by resourcefulness, and prompted by poverty, which is distinct from the imaginary imposed by the monetization of neighborhoods, a prevailing objective in urban development."[25] In this way, "rasquachification messes with the white spatial imaginary and offers up another symbolic culture—combinatory, used and reused. . . . Rasquachification challenges America's deep racial divide through acts of ultravisibility undertaken by those rendered invisible by the dominant ideology of whiteness."[26] This ultravisibility through repurposed everyday objects cultivates a sense of place by keeping cultural and location-specific memories, stories, and practices alive through a practice of what Bedoya calls place-keeping. Confronted with a white spatial imaginary, place-keeping through rasquachismo aesthetics speaks from a stance and style of resistance and adaptability through acts of ultravisibility.

Designing the creek project to create character areas defined by resourcefulness out of poverty, the creative reuse of everyday objects, and a mindfulness about stances and styles of ultravisibility are significant sources of place-keeping. They are also knowledges about a transitional climate adaptation not rooted in the colonial imaginaries of whiteness.[27] Consider the importance of this emergence for contemporary San Antonio: the ability of local government agencies to foster downtown Latino Urbanist projects is a legacy of the city council's transformation discussed in the last chapter, and more broadly, a legacy of the steady inclusion of Mexicana/o and Latina/o-origin people into positions of power. Whereas in the two previous chapters local government was a barrier to participation, now local government acts more like a bridge when it hires design firms like Muñoz and Company to bring Latino Urbanism into full public expression. It is a testament to the fight against colonial oppression that San Antonio would invest money into these public waterway development projects and bring together ecological restorations and syncretic cultural expressions as local responses to the effects of colonialism and extreme flooding. This is Escobar's design for the pluriverse. The initial design marks resistance and excess constituted by subaltern subjects at the locus of colonial difference, and it challenges dominant modernity's negation of other worlds' difference through epistemically de-linking from Euro-American sources of knowledge.[28] The creek is thus dramatized as sentient beings, sacred entities, and emplaced characters, not merely objects or subjects of "belief."[29] These transition designs for the pluriverse anticipate extreme weather to broach the question of transcultural innovation and technoscientific multitudes through cultural and ecological designs of place-keeping—all within the context of the external colonialism of global capital and the internal colonialism of public participation.

In order to substantiate these claims, the next section engages in a detailed rhetorical analysis of the topologies and tropologies of authenticity, place-keeping, and ecology in the fraught public discourse around the initial designs of the SPCP project. Throughout I will argue that while public appeals to the ecological can reify a white spatial imaginary, it is in empowering artists like Muñoz and others through local government that keep practices of place-keeping alive.

both / and.

AUTHENTICITY, PLACE-KEEPING, AND RESTORATION ECOLOGY AS "THE FICTIONAL WOMB OF HISTORY FOR COMMUNITIES IN TURMOIL"

Place-making projects like the San Pedro Creek Project are typically fraught with multiple rhetorical threads about what authenticity means, and ultimately some version of authenticity must be authorized and legitimated through various public actors, authorities, and institutions. In this way, rhetorical claims to authenticity are a site for power struggle and thus quite pro- ductive for public and political rhetorics seeking transitions in the face of ecological and social crisis. With roots in the mid-fourteenth century, the word, authentic, is based in the Greek word *authentikos* from *authentes*, meaning "original, genuine, principal" from *autos* or "self" plus *hentes* "doer, being" with further roots in "to accomplish, achieve." Thus, it is in the ontological notion of a self-doer or a self-being that we associate authentic with original, real, authoritative, and factual.[30] It comes as no surprise then that authenticity is deeply implicated in cultural essentialism and a static version of identity politics that terms a cultural expression authentic because "it has always been that way."[31] Authenticity does violence to movement and change in both creative cultural expression and ecological systems seeking to restore a landscape back to some socially desirable state. Indeed, some of the better work to emerge out of the intersections of rhetorical studies and decoloniality makes a similar argument about the dangers of presuming a fixity among identity and location in studies of non-Western difference because it moors them to cultural essentialism. In writing on identity and location from decolonial standpoints, José Cortez explains, "It is equally possible that, instead, these reflections indicate the waning of all stable borders, subjectivities, and locations . . . [and] . . . all such formalizations—the assumed relations between location, subjectivity, and the formal properties of racial identity—were only ever historical fictions in the first place."[32] Indeed, in the present context of radical transnationality, reinstituting cultural essentialism through static frameworks of identity runs the risk of irrelevancy in a world where cultural mixing is an everyday phenomenon. In the words of Mexicano/Chicano performance artist Guillermo Gómez-Peña, "This tendency to overstate difference, and the unwillingness to change or exchange, is a product of communities in turmoil who, as an antidote to the present confusion, have chosen to retreat to the fictional womb of their own separate histories."[33] This is true not just for borderlands rhetorics but for all kinds of development and ecological rhetorics still rooted in one-world ways of thinking.

The initial approval of Muñoz and Company's 40 percent design of the San Pedro Creek Project in April 2015 inaugurated a constitutive public rhetoric where claims to authenticity took center stage through various forums such as public meetings, opinion surveys, video testimonials, public opinion letters, advocacy journalism, and more. Indeed, Muñoz and Company had their own vision of authenticity: "We imagine telling the story of San Pedro Creek as a codex; a layered timeline of historical context articulated through visual art and craft unique to San Antonio. By using [these] techniques . . . we retain an authentic voice."[34] For Muñoz, cultural authenticity and design integrity means locating the origins of place-making in Latinx aesthetics as a contrast to the Spanish colonial design of San Antonio's well-known River Walk. But the initial public comments on Muñoz's 40 percent design repeatedly pushed back against the ultravisibility of the project. In a public survey (n = 113) taken at the end of May 2015 after an open house, public comments revealed that people wanted "more natural design features" and "a focus on water and green space."[35] In addition to concerns about safety, parking, and lighting, respondents were concerned about the "theme park feel." In a meeting with the county, a chair of the subcommittee reported that some publics thought "some features are 'overdone' because the illustrations are too exuberant and the colors are too vivid."[36] Other comments from the 40 percent survey expressed concerns about rising property values, gentrification, and low-impact development. But the initial takeaway for Muñoz's architectural firm was that if 80 percent of people liked the design and 20 percent disfavored it, that constituted a solid design check similar to what the River Walk experienced at a similar stage.

Throughout these public deliberations no concern was more loudly and intensely expressed than the charge of inauthenticity. In a letter to the SPCP project manager, later made public because it represented many public concerns, a local business owner on the creek critiqued the 40 percent design as "expressing an inappropriate grandiosity that does nothing to honor the history of San Pedro Creek, nor the generations of San Antonians who have lived and worked on or near its banks."[37] Critics believed this grandiosity distracted and cheapened the history and natural beauty of the creek because of its "showiness" and "flamboyance."[38] This sense of false decorum about the ultravisible design was further supported by Hector Cardenas—a figure who was instrumental in the preservation of the historic San Pedro Springs Park. He stood in front of the committee, thanked everyone for their work, and said he was glad that "things are going to be toned down."[39] This color-laden language of inauthenticity along with the preference for more natural design features didn't go over well with the design team and the project manager. They argued that "the team considered a cultural design and understands it is not the prevailing mainstream aesthetic."[40] Project manager Suzanne Scott defended the initial design renderings by arguing that the public will have to realize that "the designs on this project are quite a bit different than what people are used to. . . . It's a bolder design response. The design goals are different. This project will have its own unique characteristics that reflect the community and the culture and history of San Pedro Creek."[41] In short, supporters of Muñoz's design praised the boldness and the goals of the project based in Latino Urbanism, while critics wanted a quieter, more natural and environmental design that didn't look like "the River Walk in the holiday season, seemingly designed to attract visitors rather than locals."[42]

Meanwhile, powerful political defenders of the design team, most notably longtime county commissioner Paul Elizondo, a Westside native, claimed the design critiques based on inauthenticity were racially motivated and demeaning to the local Mexican American and Latinx communities:

"All of a sudden there is a project in the near-Westside and it needs to be critiqued. 'You've got locals doing it so you need to be guided on what fits and what doesn't fit.' *It's patronizing*," he said Friday. "It's *got a certain color to it, a Westside color*. It's not the same as the River Reach or the Museum Reach. We want the creek to bring economic development, but we also want it to be *a place for people on this side of the city and all of the city*. We have unique neighborhoods in San Antonio that were wiped out by urban renewal and were wiped out by the freeway. You get all of these snide remarks like it is too Mexican, it's too Latin. *Heck, that's what it was*."[43]

Like the design team, Elizondo locates authentic claims to "what it was" in the origins of Westside color as it was before the Texas Modern—Mexicana/o,

Chicana/o, and authentically, Westside San Antonio. For Elizondo and his supporters, these circulating rhetorics of inauthenticity were a thin disguise for racial bias against Chicana/o cultural aesthetics like Latino Urbanism and rasquachismo, and indeed a bias against Muñoz's vision for creating "a place for people on this side of the city." In the circulating public rhetorics about whether this "Westside color" was authentic or not, we can hear echoes of Gómez-Peña's critique of authenticity as a retreat to the fictional womb of separate histories. The characteristic views of the local business owner and Elizondo leverage authenticity as an unwillingness to change or exchange around their divergent points of origin—a precontact San Pedro Creek with emphasis on water, green space, and nature; and a preurban renewal San Pedro Creek with unique barrios. Negotiating this racial tension among perceptions of inauthenticity was not going to be solved by telling the San Antonio public that they need to get used to a bolder and different design. Navigating this conflict required topological folds in these authentic/inauthentic binaries, and in this case, it was translocal actors with enough cultural proximity to negotiate difference, mediate exchanges, and invent new arguments, who were able to legitimize intersecting cultural histories of San Pedro Creek today.

The agency of translocal actors in part comes from their differential positions as insiders/outsiders constantly negotiating and legitimating local/global histories, and thus it's particularly important to attend to their rhetorical performances. The first I will address was Robert Hammond, a native San Antonian living in New York City, who cofounded "Friends of the High Line"—a local citizen group who played a major role in the design of New York City's celebrated High Line Park. In anticipation of his public event, the local independent news ran an article in which Hammond argued for the integration of a landscape architect into the design process as a way to achieve authenticity:

> Build an *authentic design* and people will populate it. You don't need *gimmicks* to attract them. . . . You don't always need something like the Tree of Life to attract people. People don't go to the High Line because of the architectural experience, they go there because it gives people *a new way of seeing and experiencing the city*. . . . You don't want people to say the neglected part of town is getting the cheap version of the River Walk. Build *an authentic design* and people will populate it.[44]

Implicitly or not, Hammond's critique of Latino Urbanism and rasquachismo as "gimmicky" and "cheap" may have revealed a lack of understanding of, or distaste for, Chicana/o aesthetics. Nonetheless, his notion of integrating a landscape architect into the process gained a lot of attention from the supporters of a more natural design. That September, the project manager announced a delay in approval for the 70 percent design phase, and they made major changes, including the hiring of a renowned landscape architect. So, the procedural dimensions of public rhetoric—via surveys, letters, and public events—had effectively persuaded the project manager to assent to some of the claims for authenticity via a natural landscape. The question now was how to embody these changes to foster a shared sense of legitimacy in the design of the project. Of course, much depended on who the project managers decided to hire as their landscape architect to develop that new way of seeing and experiencing the city.

By October 2015, the project managers hired Mexico City–based landscape architect Mario Schjetnan Garduño, with the firm Grupo de Diseño Urbano (GDU), to consult on the San Pedro Creek Project. GDU is a choice that threads the needle on this crisis of public legitimacy. On the one hand, as a landscape architect, Garduño appeases critics who want the ultravisibility toned down in favor of natural design features; on the other hand, as a Mexican firm, GDU has enough cultural proximity to the Mexican American cultural identity of San Antonio that it helps the design team "best [represent] the distinctive character of San Antonio's Mexican and Mexican-American influenced environment in the creation of what will be a linear park like nowhere else in the world."[45] But yet again, the cultural proximity of a Mexican national could potentially overshadow the Chicana/o imaginary based in a resourcefulness prompted by the experience of poverty through US colonialization. In this regard, a practice of place-keeping would have to survive the diminishment of rasquachification and a global spatial imaginary that monetizes the surrounding neighborhoods and potentially entrenches racial division through gentrification as another form of colonialization. Despite these problems, the project manager told Muñoz and Company to hire and work with Garduño as their new landscape architect.

Garduño's performance of this negotiation is significant for notions of place-keeping in progressive visions of climate adaptation because of the way he authorizes Chicana/o cultural designs through a reenchantment with local ecologies, and especially San Antonio's precious source of water. For example, in his presentation to the county, Garduño drew explicit parallels between the natural history of San Pedro Creek and Muñoz and Company's newly conceived design. He showed images of San Pedro Springs as a wetlands in the 1860s and as an artesian spring swimming pool in the 1930s. Garduño emphasized that like San Pedro Springs Park, the redesigned creek would have significant interaction between people and water by adding to the design "*manantiales* (artesian springs), *ahuehuetes* (gigantic native bald cypress trees), local limestone, and the *ciénega* (wetland) to create another type of experience along the creek."[46] Muñoz's Chicana/o influence was retained in the colorful mosaic tile for the walls, fountains, and benches along the creek. But as Garduño said, "The symbolic or didactic qualities, like the Tree of Life, will serve as an inspiration from our traditions, but not copy it."[47] In short, what Garduño and GDU brought to the project was an emphasis on integration and interactions between moving water, native plants, strong trees, and pedestrian features (pathways, benches) that reflected a community-oriented linear park. As one local journalist put it, "The result is a design that is like the city and its people, deeply influenced by its Mexican roots, its Mexican-American history and culture, and its spare Southwestern landscape of stone, earth, trees, native grasses, and precious water."[48] This transitional design is what climate adaptation projects look like today in San Antonio—a thinking and practicing with speculative weather events to recreate habitats, reenchant sacred cultural places, and yet, also reify white spatial imaginaries for infill developments where global capital can comfortably invest (see figure 6).

As a project that attempts a pluriversal design—one where the water, its species, and its mythologies have a voice in the process at the very least through their entanglements with humans—the SPCP demonstrates the radical indeterminacies at the heart of any notion of an authentic design. Indeed, any attempt to locate authenticity is only ever someone's fixation of a specific space, time, and people. There are only a plurality of authenticities, and the built version is just one expression of their arrangements. The design that is built communicates which world among worlds is most privileged now, most valued today, and thus, design is a reflection on the obsessions of a contemporary moment rather than any original histories. In this way, authenticity rhetorics are tropological mirrors. They reflect and refigure contemporary power-laden social orders. Ultimately, the version of the SPCP that was legitimated by the public participation process had the effect of eroding Muñoz's design. Thus, in this version of blending local ecology with Chicana/o cultural aesthetics, the project design whitewashed a primary source of San Antonio's originality—its long and deep tradition of Chicana/o cultural practices that challenge San Antonio's white spatial imaginary through acts of ultravisibility.[49] In this view, the public deliberations around integrating local ecology and natural landscape features were an attempt to reestablish a white spatial imaginary that brings comfort to global capital investments. On the other hand, many elements of Muñoz's original design remain intact, and he retained creative control over the project design (as long as it integrated more landscape features). As of this writing, there are still four more character-areas to be designed and built, and thus,

Figure 6. Images
from Phase 1 of the
San Pedro Creek
Project. Photographs
by Kenneth Walker.

Figure 7. *De Todos Caminos Somos Todos Uno*, mural by Adriana Garcia. Photographs by Kenneth Walker.

there will be many opportunities for these deliberations around authenticity to change the ratios among Mexicana/o, Chicana/o, and southwestern landscape influences. So, despite the whitewashing, the ability of local government to hire, retain, and defend artists like Muñoz speaks volumes about the importance of place-keeping and the power Chicana/os have achieved in San Antonio's city politics.

A final performance that communicates both the limits and possibilities of place-keeping in pluriversal adaptation projects is represented through one of its award-winning public mural projects. *De Todos Caminos Somos Todos Uno* (*From All Roads, We Are One*), created by San Antonio artist Adriana Garcia, depicts contact between the Indigenous Payaya/Coahuiltecan people and the Spaniards through time and to the present day. As in the SPCP itself, in the mural, water is the gathering place for all aspects of life—from the Indigenous symbols of the anhinga water bird and the blue jaguar river spirit at each end of the mural; to the ancestral faces of a hunting and gathering peoples and a literary and colonizing peoples; to a synthesis of youth floating in the creek and young families playing by its trees and in its water (figure 7). The mural is a cyclical and generational narrative of place-keeping in deep time where the material forces of the environment foster the conditions of cross-cultural and cross-species contact over generations of sustained liveliness in San Antonio. In Garcia's depiction, it is the supernatural qualities of water, trees, and animals that provide guidance and direction for

[handwritten in left margin:] ok but what about violence & compromise?

generations of living well together. As a creative act that honors and recognizes what animates the good life, the mural is simultaneously an engagement of other worlds' difference at the site of the colonial wound and a challenge to one-world development. Thus, while Garcia's mural is certainly not rasquachismo, it can be viewed as a de/colonial creative expression of a world among worlds, an ongoing contact on the verge of yet another generational transformation guided by the ancestral blue jaguar and water bird spirits. The mural is thus a monument to creative acts of pluriversal place-keeping where cultural and location-specific memories, stories, and practices are continually retold and redone anew as rhetorical folds of the many become one. The mural perpetually asks, who is the we gathered here? And from what roads do we all become one at this place?

HOUSING JUSTICE AS RHETORICAL INHABITATION IN PLURIVERSAL CLIMATE POLITICS

The internal colonialism of authenticity rhetorics and the white spatial imaginaries of ecological restoration certainly shaped the final design of the first phase of the SPCP. Yet, through the advocacy of a protected enclave of powerful actors, significant sources of place-keeping remain. In place-based projects with pluriversal designs for transitional adaptation, one should expect these kinds of local struggles and learn from them to animate the difficult choices of place-keeping projects. But, when the SPCP is placed back into the context of the global growth machine, how can local actors hope to shape the pathways of development rhetorics that demand the creation of consumer markets? How will the demands of global capital investment affect transition discourses as practices of pluriversal place-keeping? This section engages these questions through an analysis of housing justice as rhetorical inhabitation.

Directly behind Adriana Garcia's mural is the Soapworks and Towne Center, an apartment complex that used to be a soap-making facility in the mid-nineteenth century but today provides one of the last affordable housing units in downtown San Antonio. As the San Pedro Creek Project neared completion, a Houston-based global real estate firm bought the properties, began renovations, and increased rents to a middle range. By April 2018, over sixty apartments had been vacated by people who could no longer afford them. Other residents facing displacement created a tenant's union to protest the renovations; they argued the city should buy the complex and offer it as affordable public housing units.[50] Indeed, the irony of these developments were not lost on housing justice advocates—directly behind a mural depicting relatively peaceful colonial contact was a site of the recolonization of working people.

From the perspective of the Soapworks residents, because it was city and county policies that led to the redevelopment of the creek, it was therefore local government's responsibility to offset the displacement of working-class residents at Soapworks, the majority of whom were people of color. These residents, along with housing justice advocates, pointed out that displacement happens all too often when the city redevelops along its watersheds. An antecedent case was the displacement of three hundred residents at the Mission Trails mobile home park where local government rezoned and redeveloped the San Antonio River Walk along the missions, bulldozed the trailer park, and replaced it with a luxury condo complex.[51] A grassroots housing justice group called the Vecinos de Mission Trails documented this displacement with fifty-one interviews from residents of the trailer park.[52] In their executive summary, Vecinos wrote a familiar narrative: Residents were predominantly Mexicano/Mexican American (85.5 percent), low or very low income, Spanish-speaking, young (44.9 percent were under eighteen), and many were immigrants (20 percent of residents).[53] The main shocks to these long-term and mostly trailer-owning residents came in the form of housing insecurity (multiple moves, overcrowding, or homelessness), declining health, economic insecurity, and negative impacts on children.[54] Others noted how quickly the displacement happened, that it seemed to be for nothing, and that they felt betrayed or not listened to by city leaders. Ultimately, what Vecinos de Mission Trails recommended was for state and local governments to facilitate home ownership by giving homeowners the in- right of refusal for rezoning/sale of property, for cities to provide capital and technical assistance for residents to acquire and manage their own parks, and to take on displaced residents as policy advisors. Indeed, the residents of Soapworks and the housing activists who continually fight for them point to the Vecinos de Mission Trails as a reason why the city should purchase the Soapworks complex and establish it as a model of affordable public housing.

Like the Vecinos de Mission Trails case, the Soapworks complex demonstrates a central tension between local people and global capital that I want to address here through rhetorically inflected notions of inhabitation. To their credit, even though it was the mayor and city council's job to attract global capital investment into downtown, they also held on to a modicum of what rhetorical and critical scholars have called inhabitance. Based in the work of French Marxist theorist Henri Lefebvre, developed by Mark Purcell, and recently applied to the Vecinos de Mission Trails case by Marisol Cortez, inhabitance is a theoretical basis for rights claims arising from the limitations of citizenship, and the need to shift urban decision-making away from capital and the state and into the hands of residents and all those who inhabit city spaces. In this view, "those who inhabit the city have a right to the city." Thus, rights can be defined through participation—that is, those who use those spaces everyday have a right to participate in decisions about how to develop that space and prioritize its use value over its potential exchange value.[55] In prioritizing participation and the use value of inhabitants, Cortez draws on the Global South Indigenous concept of *buen vivir* to argue that a decolonial sense of inhabitation would value nature with rights itself: "The right to the city in fact depends upon the rights of nature, ultimately folding river health and integrity into an expanded human right to inhabit urban space well."[56]

Rhetorical scholars have also theorized inhabitation as a metaphorical (tropological) and material (topological) practice of bordering and as a more-than-human activity with provocations for inventional practices.[57] Most recently, in calling for further rhetorical research on place, Thomas Rickert has argued that environments inhabit us just as we inhabit them, and this acknowledgment calls for a broad concept of inhabitation where physical spaces and sensory information are participants in rhetorical practice.[58] Such views might also be aligned with pluriversal designs for transitional adaptation projects such as the SPCP. As Cortez notes, the crucial practices of inhabitation are ones where ecological health shares a strong relation to the health of humans already inhabiting a place. Disturbance in modern-colonial structures like drainage ditches are thus opportunities for reconceiving the good life among the ruins of capitalism.[59]

More rhetorically, such disturbance ecologies might foster ethical practices of inhabitation that fold topologically as both becoming informed about
a place (its commonplaces) and further informing a place through rhetorical practice (its un/expected turns). In Casey Boyle's recent extension of
Muckelbauer, this notion of rhetorical inhabitation also depends on sharing
that inhabitation with others by connecting through the immediacy of being there (virtually, in person, or in spirit). Practices of place-keeping and
rhetorical inhabitation thus function transductively across spatial temporalities and always toward place as a mode of invention.[60] In what Boyle calls a
transversal practice, inhabitation is not just emplaced and common but also
placed again with new information.

For the SPCP, the efforts to ethically recreate and restore the inhabitance
of many species of grasses, birds, invertebrates, trees, and more must also
work to keep in place many working-class inhabitants for the purposes of
place-keeping under the constraints of gentrification. It matters what hu-
man/nonhuman assemblage is there, what/who is remembered, and what/
who is informed, and what/who informs inhabitation. In their relation to
global capital, pluriversal and rhetorical concepts of inhabitation cannot be
practiced through disturbance ecologies to create nature playgrounds for the
well-to-do alone. Inhabitation in the Soapworks case would mean including current working-class residents in the decisions about the privately run
property, and even prioritizing a reasonable quality of life (use value) rather
than maximizing profits (exchange value). As pluriversal rhetorical concepts, inhabitation and place-keeping cannot be only bound to current liberal humanist legal structures. These are excessive concepts that re-create
worlds of difference at the fractured locus of the colonial wound, and they
therefore challenge dominant modern assumptions about the priorities of
the global growth machine. Such discourses and practices are then provocations to rethink inhabitation as a reconsideration of the relations among
worlds for each specific context and in anticipation of the next disturbance
via extreme weather events.

To its credit San Antonio's mayor and city officials vowed that cases of
blatant housing injustice like the Mission Trails would never happen again.
With an appropriate push from local housing activists, the city's response
to the Soapworks displacement demonstrates both the potentials and limits
of inhabitation policies that at root seek to constrain the profit-maximizing
freedoms of developers. First, the mayor created a housing policy task force,
which included housing justice advocates, to study displacement and offer
policy recommendations. Meanwhile, multiple local officials committed to
finance housing assistance for anyone who was displaced by the SPCP. As

the housing policy task force worked to finalize their recommendations, many of the recommendations were already implemented in the city's 2019 budget, which tripled the funding for housing initiatives ($25 million) and raised the living wage to $15/hour. The mayor called this a "back to basics" budget that "furthers our commitment to expanding affordable housing by approving $25 million towards developing a coordinated housing system and providing new resources for rehabilitation to preserve affordable housing."[61]

But city-level housing policies based on inhabitation cannot simply exercise guilt by throwing money at the problem. The additional policy recommendations from the housing task force use inhabitation-based policies to channel global capital and improve the lives of current inhabitants. They recommended "by-right zoning," which would allow housing projects to receive automatic zoning approval if half of their units are affordable; larger incentives for developers to build low-income housing; allowing city bond money to be used for public housing; increasing funding for payment assistance and housing rehabilitation programs; and increasing public accountability by creating new positions within the housing and neighborhood services department, including an executive position within the city managers' office.[62] Thus, through city policies based on principles of inhabitation, projects that anticipate climate crises also need to anticipate the social crises that arise from free market urban development. If climate adaptation is to be more than a function of a reaction, and instead a function of pluriversal transitions, policies based on inhabitation that intervene into dominant modes of political economy must be paramount. To paraphrase Jessica Guerrero, the director of Vecinos de Mission Trails, cities addressing notions of a just transition in climate adaptation will also have to mitigate the "natural" disasters the city itself creates.

From a broader perspective, the relationship between place-keeping and inhabitation in just climate adaptation projects cannot only be reactive; they must also be reconstructive of the pluriversal. As Mark Pelling has shown, there is potential in thinking about climate adaptation as a process of political transitions and transformations.[63] In this case, the discourses and practices of place-keeping and inhabitation kept the SPCP appropriately political so that at least some measure of a just transition was possible. Place-keeping and rhetorical inhabitation helped provide these transitions by achieving some measure of transition from urban disenchantment to urban re-enchantment, from displacement to re-placement, from economic segregation to economic integration. Without a doubt, these fights will continue, as the city is currently pushing to hire a qualified housing executive committed to practices of inhabitation and committed to bringing those under the threat of displacement into the role of policy advisors. Such procedural justice moves could help articulate revitalization among creek inhabitants as one version of a pluriversal design for transitional climate adaptation with a more inclusive vision of the places in which we work, live, play, and worship. As the next four phases of its development continue, the SPCP will have much to learn from grassroots movements for climate and housing justice in San Antonio, if it is to hold on to any sense of a just and pluriversal transition that can face up to the historical and contemporary legacies of this three-hundred-year-old colonial city.

CLIMATE POLITICS AT THE LIMITS OF PLURIVERSAL ADAPTATION

The intersecting issues of climate and housing are creating nothing short of an emergent coalition for intersectional climate justice in San Antonio today. The San Pedro Creek Project could place San Antonio on the forefront of conversations about what pluriversal designs based in place-keeping and inhabitation can mean in transitional climate adaptation projects. Local efforts to stop the displacement of working-class residents will clearly not be a silver bullet to the problems of urban development, sustainability, and social reform, especially with a future of abundant extreme weather events. Few, if any, cities in history have grown by a million people in just a few decades and appropriately handled gentrification. As many locals like to point out, when Austin faced similar issues it did little to help its working people. While San Antonio is differently located, and for now, growing at a slower pace than Austin, it does have examples like Austin to learn from. Perhaps equivalence topologies will not be the dominant development rhetorics in San Antonio.[64] In its struggles with the indeterminacies of authenticity and inhabitation, one witnesses quite a divergent rhetorical practice, underwritten by divergent subjectivities now authorized through the protected enclaves of local government. These achievements are surely the product of a long struggle to include and empower marginalized people in the mechanisms of local democratic governance. But to what extent these recent achievements will face more resistance from global capital and whitewashing from local actors invested in historical structures of privilege and security remains to be seen.

For now, and in the near future, the success of these incremental steps toward a just and pluriversal transition in climate adaptation projects will have to confront a different kind of colonial monster in the coming generation of supercharged climate events. As a place with a multigenerational memory of climate adaptation through conditions of poverty, San Antonio does foster a space of resistance to dominant narratives of modernity/coloniality, as the epigraph that began this chapter notes. If these memories are preserved and implemented into policy, San Antonio could be on the threshold of what it means to build adaptive capacities and develop a broader sense of a just transition for all inhabitants who will live together through the climate politics to come. A pluriversal climate adaptation will achieve something like sympoiesis in its trading displacement for replacement, place-making for place-keeping, and the disenchantments of modernizing for the reenchantments of ecologizing.[65] This speculative vision is perhaps a best-case scenario—one where San Antonio continues to do the right things that anticipate the speed and scale of climate influences on a planet now just over one degree warmer due to human pollution. This is the planet we live on now, but it is not the planet my son and daughter and all earthly inhabitants of the next generation will inherit. That planet will likely be over three degrees hotter by midcentury, even if the global economy is held to the strictures of the Paris Agreement on climate. Even that outcome now is a point on the range of best-case scenarios with an entire bell curve of worse-case scenarios.[66] On that planet, transitional climate adaptation projects will be beside the point. If cities like San Antonio, and national governments across the world, cannot fully engage in antiextractivist and pluriversal climate politics right now, all inhabitants may be left grasping for earthly survival.

PART III
Speculative Pluriversal Rhetorics

CHAPTER FIVE

The Rhetorical Auguries of Climate Politics

The Assaults will not be discrete—this is another climate delusion.
Instead, they will produce a new kind of cascading violence, waterfalls
and avalanches of devastation, the planet pummeled again and again,
with increasing intensity and in ways that build on each other and
undermine our ability to respond, uprooting much of the landscape
we have taken for granted, for centuries, as the stable foundation on
which we walk, build homes and highways, shepherd our children
through schools and into adulthood under the promise of safety.

DAVID WALLACE-WELLS

Here we find the habitual vice of epistemology, which consists in
attributing to intellectual deficits something that is quite simply a
deficit in shared practice.

BRUNO LATOUR

And that had been fine for a while, until once again she felt called to
incarnate the theory she taught, to realize ideas in action, to widen
praxis to the size of the love coursing through her.

MARISOL CORTEZ

IN JANUARY 2019, JUST A year after celebrating its three hundredth year as
a colonial city, and two years after joining the Paris Agreement, San Antonio
launched the development of its first Climate Action and Adaptation Plan
(CAAP). For over a year, and with expensive external consultants, a diverse
coalition of ninety leaders from across the city's sectors created the first draft
of a plan that committed San Antonio to net zero greenhouse gas emissions
by 2050. After much public discussion, and a vocal concern from the business community about the costs associated with a full transition to renew-
ables, the city council delayed a vote on the plan until elections had passed.
In the meantime, the city-owned energy company, CPS Energy, secretly

lobbied for changes to the plan and refused to sign on until its demands were met. When the second draft was released, the plan kept the abstract goal of net carbon neutrality by 2050, but eliminated specific commitments to cut emissions from CPS's power plants, to fully transition to carbon-free vehicles on San Antonio's roads, and to mandate a reduction of citywide energy use in buildings by 40 percent by 2040. In short, the second draft of the CAAP was gutted of any real change to business as usual. In a real display of magical thinking, at one point CPS declared they would both support the CAAP's carbon neutrality goals and burn coal well into the 2060s. The question that lingered for many San Antonians was if CPS would not listen to its people and their representatives on this issue, do the people really own the city's energy company?

City-level plans like San Antonio's CAAP are significant vision documents—speculative road maps, future guides—for difficult political choices that are perpetually prospective. As speculative visions of practice, such plans are provocative platforms for pluriversal rhetorical practice in the near future. Of course, city politicians love vision documents because they keep difficult political choices at a distance while simultaneously positioning politicians as a version of Lady Liberty leading her people through difficulties to a bright, prosperous, and equitable future. With classic tropes of modern/colonial progress, city-level planning documents reify the image of a shining city on a hill, that exceptional beacon of hope courting capital through legal permissions to extract and accumulate, inevitably at the expense of others. Since approximately 2009, many major US cities have adopted some version of a climate action and adaptation plan in anticipation of local climate crashes, and in hopes of creating resilient cities. But often their hidden histories and contemporary realities belie their public prophecies. What I aim to accomplish in this final chapter is to engage in a reading of this city-level vision document, not to trace its rhetorical development, but to theorize place, power, and politics as a land struggle for a city's long unfulfilled desires for equity in the new context of climate breakdown.[1] Meanwhile, I hope to engage this planning document in order to identify a few near-term futures of pluriversal rhetorics.

San Antonio has a somewhat unique version of a climate action and adaptation plan because at its center was a rhetorical struggle around the concept of "climate equity." In theory, the CAAP made climate equity a central principle in its design and implementation. In practice, this meant the CAAP created a Climate Equity Technical Working Group to help guide its visions, and in turn this team created an equity screening mechanism that would "screen" all other policy choices through the lens of equity. In these ways, climate equity could help guide the city's version of progress. Of course, the irony of planning for equity deep in the heart of one of the most inequitable cities in the United States was not lost on most skeptical San Antonio observers. After all, the number of long-forgotten city-level plans that have promised some version of equity are too many to remember. In theory, efforts like the climate equity working group and equity screening mechanisms could hold decision makers accountable for acknowledging historical legacies of colonialism and forcing fair decisions during implementation. But as we have seen throughout this book, equity is not a single thing to be possessed. If indeed equity is based on one-world cosmologies of possession, it becomes mired in perpetual struggle where equity could just as easily backslide as it can progress. This is the crucifix of equity—one version of equity is sacrificed while another version is born. Equity has long been a crucifix for reifying city- and state-sponsored practices of violent segregation. So, it's worth asking how climate action plans can foster pluriversal forms of equity—not as possessions but as right relations to land and people through interdependence in the difficult work of living together and not without nonhuman others.

Throughout this book I have argued that local histories of climate politics are relevant for local futures. And what I have learned through San Antonio's local histories is that equity-based discourses and practices have only been forced on the city whose primary practice has been progress and protection for its colonial legacies. For example, it took San Antonio until the 1970s to place indoor plumbing in the Westside barrios, fifty years after the devastation of the 1921 flood, and fifty years after significant public demands for equitable development from Westside residents and local scientists. Before that time, most inhabitants used cesspools or septic tanks for sanitation. The legacy of San Antonio's flood protection, crystallized by Olmos Dam, is an eighty-year legacy of city-driven inequity. Even the powerful secondwave Chicana/o organizations bowed to capital's hostage tactics of creating barely living-wage jobs or leaving town. Even clear-eyed equity-based adaptation projects like the San Pedro Creek Project faced constant assumptions of white-spatial imaginaries and their assumed economic benefits that in effect eroded Mexicana/o and Chicana/o cultural ways of being and practicing. In San Antonio, equity is a tricentennial haunting. If indeed the primary problem with climate change is the ability to foster an appropriate political response, then the colonial histories that have created this monstrous situation must be radically questioned. But rhetorical questions are not only about today's histories. They are also inventional questions about what is coming and what other practices must emerge to create a more pluriversal sense of political possibility under climate breakdown.

For too long now, climate scientists have been telling publics that the force of climate breakdown will be a central organizing principle for most aspects of life in the twenty-first century. If the local histories of twentieth-century climate politics in San Antonio are any gauge, it tells us that the slow crawl of political incrementalism and cultural inclusion is completely misaligned with the severity, scope, and speed of the coming climate effects. It simply cannot take San Antonio another thirty years to cut coal from its power plants. It simply cannot take San Antonio another thirty years to dispel its illusions of inclusion.[2] These few decades at the beginning of the new century demand a new response, new practices to simultaneously think deep and move fast together through divergence and diplomacy. This, or the alternatives will be that a one-world dominance cannot adapt, it will not be resilient, and we will all be radically vulnerable to the supercharged forces of a human-altered climate. Cities like San Antonio have a small window of opportunity to ask themselves now what sacrifices they're willing to make to achieve the kind of responsivity publics will need to simultaneously face the realities of climate change and break through the illusions of inclusion. Otherwise, other forces will make those decisions, and political change will forever play catch-up with the new sovereign that is climate change.[3]

This final chapter uses San Antonio's CAAP to speculate on futures of rhetorical theory and practice through two overarching questions: (1) What is on the horizon for local climate politics in the next generation? And (2) What kind of heuristic and hermeneutic can rhetoric provide for invention and interventions into these climate politics? These two questions address the near future of climate politics broadly but through specific city-level planning documents that speak to contemporary moments of local climate adaptation. The second question assumes that climate politics will become a ubiquitous aspect of rhetorical studies in a similar way that digital rhetorics now permeates most rhetorical study (indeed, the digital and the climatic is another critical intersection that must be constantly incorporated into each project as the rhetorical work on data centers and technological mediation of weather shows).[4] But whatever the theoretical and methodological investments—critical, cultural, rhetorical, engaged, ethnographic, historical, interdisciplinary—climate politics will increasingly become a ubiquitous rhetorical power that will need to be theorized and put into critical practice. To begin to address just a few of these intersections, and after a brief section on San Antonio's version of climate equity, the sections that follow on pluriversal rhetorical science studies, migrations, energy politics, and extreme weather disturbance begin from an analysis of San Antonio's CAAP in the context of the latest climate science. Ultimately, I argue that these shared areas of study and more will also oblige a shared praxis across subareas of rhetorical scholarship that will continue to build on un/

common places and take un/expected turns. As rhetorical scholarship of all kinds grapples with the forces of a changing climate, it seems appropriate to conclude with a sense of rhetoric's future potentials for making kin, fostering entangled worlds, and staying with the necessary trouble of decolonial politics.[5]

THE CRUCIFIX OF CLIMATE EQUITY

San Antonio's Climate Action and Adaptation Plan (CAAP) defines equity as "our policy-making, service delivery, and distribution of resources account for the different histories, challenges, and needs of the people we serve. Equity differs from equality, which treats everyone the same despite disparate outcomes."[6] Through this definition, the CAAP attempts to grapple with adaptation as a method of social reform to address some of the legacies of coloniality. Engaging in differential movement via climate equity is commendable, but what will futures of equity mean when climate variability, extreme weather, and climate breakdown become major political forces of the next century? A pessimistic view might hold that San Antonio's CAAP acknowledges its long history of coloniality only as a stage for a fallacious heroic performance of equity that belies a city's other purpose to extract and accumulate capital under the guise of adaptation and resilience as a contemporary form of economic growth. A more optimistic view might point to the pragmatic viability of the technical working group and the climate equity screening mechanism as frameworks for "the intentional consideration of equity issues in the implementation of CAAP strategies, i.e. policies, programs, and budget decisions."[7] Indeed, the climate equity framework was created to ensure that those most affected by climate change would also guide the city's decision-making in five areas: access and accessibility, affordability, cultural preservation, health, and safety and security.

But it is clear that the close relationship between inequality and economic growth will create a climate equity crucifix across these five categories. For example, if we read the San Pedro Creek Project (SPCP) from the last chapter onto the CAAP's climate equity screening mechanism, the SPCP has substantial benefits along the lines of access/accessibility, cultural preservation, health, and safety and security, but also, many negative effects along the lines of limiting affordable housing, increasing barriers to homeownership, increasing displacement, and increasing the overall cost burdens on working people and people of color. It is still unclear to many observers how decisions will be adjudicated when local projects are both beneficial and detrimental to a range of "equity gains." There's something more pernicious at work in calls for equity gains in the context of one of the most notoriously segregated cities now reaping the benefits of one of the largest energy and population booms in its history. That malevolent force is precisely the systemic relations of domination based on coloniality now embedded into global economic growth models that drive local inequality, especially in major cities undergoing economic transformations. The CAAP's Climate Equity Technical Working Group certainly came to this realization when after eighteen months of difficult labor to center climate equity in this city-level plan, they drafted an open letter to the mayor and city council that announced their disappointment in how its "pioneering work has been significantly watered down, and inaccurately conveyed in the current plan." In particular, the equity working group had specific concerns about how they had been relegated as an advisory subcommittee, how fair and transparent community engagement processes had been elided, and how funding for low-income communities of color had been de-prioritized. They wrote, "We understand equity to be *both* about righting historic wrongs *and* transforming present-day decision-making processes, including who gets to sit at the table. Both are essential, and both are weak in the current iteration of the CAAP."[8] Unfortunately, the working group also had trouble conceptualizing the paradox that inequality is structurally produced by global capital development models often in the name of equity, and any effort to right historic wrongs and transform decision-making would also likely require new visions of economic development.

If climate change is indeed the new political force of the next century, in ways that modernity was a century ago, one has to wonder just what a future of equity will look like when contemporary climate plans cannot even acknowledge the root of the problem in economic development models based in modernity/coloniality. Potentially, without a full reckoning of these structural issues, climate will not be a cause for equity but a cause for a new wave of neocolonialization that local climate equity mechanisms will be helpless to address. After all, the horrors of inequality and climate change are human made and are an inevitable outcome of colonial orders that dominate and permeate every aspect of contemporary life. If climate breakdown finally does create a rupture in one-world economic development models, can anyone imagine that the response to these ruptures will be a more broadly shared prosperity? Or is it rather that climate breakdown will give cities just the opposite—a new political force that ruptures one-world development models and drives inequality to its extremes?

In one of the better books about the human future of climate politics, David Wallace-Wells lays out in detail the reasons why climate politics will be the force of change in the twenty-first century just as modernity was the force of change in the twentieth century. Consider: well beyond all of human history on this planet (the last 800,000 years), there has never been as much carbon in the atmosphere as there is right now. Half of the carbon the world has put into the atmosphere was emitted in the last thirty years, and it shows no sign of stopping. This pollution has put the world on the threshold of 2 degrees Celsius of warming, a threshold that used to be a red line for avoiding the most catastrophic outcomes. Just a few years later, with no single industrial nation on track to meet its Paris commitments, "two degrees looks more like a best-case outcome . . . with an entire bell curve of more horrific possibilities extending beyond it and yet shrouded, delicately, from public view."[9] At 2 degrees of warming, the Earth's ice sheets will be in full collapse and major cities across the tropics will become unlivable with hundreds of millions suffering from water scarcity and thousands more killed by heat waves. At 2 degrees of warming, it is estimated that 150 million people will die from air pollution alone—double the death toll of World War II.[10] This is a best-case scenario. It does not even consider what humans are capable of doing to themselves.

In order to meet the 2-degree threshold, the globe will have to halve its global emissions by 2030, in roughly ten years from this writing. Today the globe is on a path to get to 4.5 degrees of warming by 2100—more than double the 2-degree threshold of catastrophe. As Wallace-Wells notes, the last time the earth was that warm, palm trees grew in the Arctic. The most recent IPCC report states that if the world acts and implements the Paris commitments, we are still likely to get a catastrophic 3.2 degrees of warming.[11] By almost any measure, including the one where humans act in a coordinated and unprecedented fashion akin to mobilizing for a world war, the next one hundred years of life on this planet is in for another mass extinction created by humans. Climate change will be the new force of production in the twenty-first century. It won't just be more hurricanes, fires, floods, and distant refugees. As the new force of production, climate will affect every aspect of human life from interpersonal relations to global treaties. Our silence on climate change now belies the centrality of climate politics for the next century where arguably climatic change will be the center of every political conversation.[12]

Against this silence, I have argued that extreme climate events will make colonial social orders based on systemic domination and exploitation radically apparent, which also allows those relations to be questioned and reassembled in radically divergent ways. The structural inequities created by colonial relations now legally protected through unfettered capitalism are seemingly intractable. If political history is incremental, and if adaptation plans are completely out of sync with the scale and scope of the problem, it is extremely difficult to see how addressing the root causes of climate in colonialism can be achieved in the next few decades. Incremental politics won't help. In this context, what possible hermeneutics and heuristics can rhetoric provide at the end of modernity's progress and for the necessary work of decolonizing, decommodifying, and decarbonizing? What will happen to public life, public institutions, and democratic governance in these new contexts of life after warming? If cities like San Antonio can barely pass watered-down climate action and adaptation plans right now, what does that mean for a future wherein we have largely underestimated the speed, scope, and severity of climate breakdown?

In his essay on democracy and climate change, Ralph Cintrón staunchly critiques all varieties of potentiality marshaled in the names of equity, rights, and freedom discourses because ultimately, they function as a set of moral possessions that one must have, which means others cannot have. Pointing to processes of individuation (self- or group-level) that are mired in perpetual competition among other individuals/groups, Cintrón writes, "This sort of atomized self, swollen with self-interest and importance, uses rights and freedoms to cultivate a vast permission to legally extract and accumulate at the expense of others."[13] In seeking to find a heuristic for an opposing semantic field to generate alternative ways of seeing oneself in community, Cintrón reads through evolutionary and ecological sciences to offer the notion of the deep commons. Reflecting on species competition, he notes, "The deep commons may not neutralize but, rather, include the forces of individuation that also underpin the capitalist narrative."[14] So, somewhat paradoxically, the deep commons includes individuation, competition, and is never fully outside of colonial/capitalistic forces. From the microworld of protoplasmic bacteria to the macroworld of biogeography, processes of individuation like private ownership, extractivism, and accumulation are co-created, co-constituted by the deep commons. As Anna Tsing would have it, capital relies on these edge spaces of the deep commons in a process she calls salvage accumulation—the amassment of wealth under capitalism by converting human/nonhuman relations into capital gains.[15] Even in the purest forms of mutualism—a form of a symbiotic relationship that also includes commensalism and parasitism— there is the need to eat in order to survive, and thus quite often, the everyday act of a sacrifice in order to consume. What is the human? To be human is to be in interspecies relationships and to live at the expense of others—a transfer of energy, a kinetics of energy—until your own payment comes due and a sacrifice is required to keep holobionts functional and wildly diverse.

It is in these interdependent and interspecies relationships that equity as shared sacrifice, as distributed prosperity, as part of ourselves, must be thought through outside the spaces of capital investments and state control, if there are any left. Reading San Antonio's climate equity through Cintrón's critique, it's easy to see how the city uses the shadow of fairness and impartiality to increase the value of its lands, to fill its city coffers, and to accumulate at the expense of others without the means to afford the ever-increasing taxes. The privatization of property, and the city's value minus its debts, is the real version of equity the city is interested in as it claims to account for the different histories, challenges, and needs of the people it serves. In using equity to accumulate and extract against its own working inhabitants, the city of San Antonio is missing the real value of a climate equity—a set of responsibilities it has in order to create relationships based in practices of reciprocity not possessions, of circulations not accumulations, and of care not apathy when civilizational collapse is on the brink. Resisting the translation of equity as fairness to equity as increasing exchange value is a fight to foster interdependent relations through public spaces, high-quality and affordable public housing, and interspecies gratitude created by shared practices of sacrifice. These forms of resistance may seem impossible, but they are worth considering if we are to account for the deficits of shared practice now threatening shared realities, shared prosperity, and this shared planet. The continual critique of unfettered capitalist markets is a must, but ultimately a more pluriversal version of equity otherwise, elsewhere, and yet somehow within the logics of state-capital control must be articulated, even if these can only be found in small moments with love, humor, and gratitude.

TOWARD PLURIVERSAL RHETORICAL SCIENCE STUDIES

At the local level of everyday rhetorical practice, I have argued that climate breakdown will create the conditions for new forms of pluriversal praxis—a proliferation of cross-species coalitions, alliances, and affinity groups that through critical rhetorical and democratic practice will confront the authoritarian and imperial impulses to invade, secure, border, and extract in the wake of climate disasters. These conditions should compel rhetorical studies broadly to recognize the roots of structural oppression in coloniality and to fully embrace coalitional work as a response at nearly every scale—in our teaching, service, research, and activism. Following the lead of Karma Chávez, Adela Licona, and María Lugones, I understand coalitional moments as turning points and historical junctures that characterize the times. As a site of partiality, coalitions are based on affinities through differences, otherness, and specificities of identity that critically matter because they are also critically limited. If modernity/coloniality is a technology of segregation, pluriversal understandings of coalitional politics characterized by divergence and diplomacy will be a necessary collective response that must also be informed by whole new visions of technosciences from below, including the necessary rhetorical work in Black feminist technoscience, Chicana feminist technoscience, Indigenous technosciences, and more.

The burden of this coalitional work cannot be placed solely on scholars who have been leading this work for over a decade now; rather, the burden of this work is on all rhetoricians seeking to articulate de/colonial theory to rhetorical science studies to make kin for coalitional politics. Despite the well-known relationships between colonialism and science, rhetoricians of science, technology, and medicine (RSTM) do not have many studies of their own to look to in this regard. Beyond a few excellent case studies of biocolonialism,[16] medical colonialism,[17] race relations in medical science,[18] and most recently, rhetorical ecologies that co-labor across trophic scales,[19] how can it be that a field as vibrant and diverse as RSTM has largely fallen silent on the issue of de/coloniality in science studies? As I argued in the first chapter, rhetorical scholars could do more to follow the lead of de/colonial science studies scholars like Sandra Harding, who in her most recent work on "One Planet, Many Sciences" argued for keeping both eyes open on contemporary Western science and many other cultures' scientific practices and legacies. As she demonstrates by building from feminisms of color, approaches to de/colonial science studies should be pluriversal in their proposals to: (1) integrate endangered Indigenous knowledge systems into the sciences of the Global North; (2) de-link from Western science altogether; (3) integrate Northern sciences into Other sciences; and/or (4) transform Northern sciences on the Southern model.[20] Yet, as Druschke has continually argued, such work is not likely to be successful unless it is done in an engaged manner that co-labors with sciences from below.

Thus, the task is an immense one. It requires bordered sites of study and validation from the communities who practice this knowledge—both for sciences from below and also for contemporary Western sciences. In these pluriversal futures of science studies, fieldwork cannot solely be constituted by working with scientists alone. Rather, the engaged fieldwork that maps Western technoscience to local publics with their own versions of technoscience is essential. This kind of rhetorical scholarship can help address the gap in shared practice that might rebuild trust in public institutions again. But whether it be fieldwork, archival work, or engaged rhetorical criticism, what seems most important is incarnating theory into action and engaging in a shared praxis based on love, conviviality, and harmony that works against colonial social orders. This book has been one attempt to invest critical cultural concepts from de/coloniality into ecological and environmental rhetorics, and on this front there is much more work to do in rhetorical (science) studies and beyond.

Yet, as I also argued in chapter 1, engaged work does not have to be the only future for rhetorical science studies. In addition to continuing the good work already being done, de/colonial histories of rhetorical science studies have yet to be fully articulated by those scholars so well positioned to do so. Rather than make hollow claims about decolonizing science studies, rhetorical scholars might listen much more carefully to those scholars already addressing issues of de/coloniality and technoscientific domains such as Damián Baca and Romeo García and Angela Haas and Erin Frost.[21] For example, García and Baca have most recently noted how postcolonial scholars have long identified the emergence of the macrohistorical rational European man of letters and science as parallel to conquest and expansionism. Moving from the work of Pratt, they use Carl Linnaeus's *Systema Naturae* as a primary example of systematizing nature under one-world frameworks with parallel effects on people. Both have the effect of placing white European scientists in a "managerial position" who rationally classify through explorations and colonizations.[22] Rhetorical science studies might take a hard look at itself by following the lead of such scholars and begin to ask questions about the histories and futures of science studies in relation to colonialism and coloniality. As Chávez has suggested for rhetorical studies more broadly, such disciplinary work is not about inclusion but about recasting what is considered rhetorical, and indeed, what is considered scientific.[23] Or as García and Baca put it, such approaches must move beyond disciplinary reform and instead engage by following the lead of decolonial scholars committed to epistemic, political, and ethical projects from elsewhere.[24] These may be some of the questions pluriversal rhetorics can begin to approach, but first, to realize its own coalitional possibilities, rhetorical science studies must also reckon with its colonial monsters and rid itself of modern/colonial one-world progress and instead foster its own version of scholarly pluriversality. This move no longer seems like an option, but a requirement.

LOSING THE PATTERN: MIGRATIONS AND NEOCOLONIAL STATES OF CONTAINMENT

In the summer of 2019, the federal government opened a new tent city in Carrizo Springs, Texas, that housed 3,000 new immigrants, half of them teenagers, and most of them seeking asylum from Central American countries. That summer migrants topped 100,000, a level not seen since 2007. Over 84,500 of these people were traveling with their families.[25] While one cannot underestimate the extent to which organized crime, systemic corruption, and violence drives current migration from Central America, the bigger picture is one that also connects displacement and migration with climate-change-induced food insecurity. *New York Times* columnist Nicholas Kristof drew a stark portrait of Guatemalan families forced to immigrate because drought and severe storms associated with climate change had made the weather so unpredictable that food no longer grows on many farms.[26]

Thus, in addition to the US complicity in the political situation in Central America since the time of Reagan and the establishment of the School of the Americas, one can now add the contributions of the United States to climate change as a driver for Central American migrations north to the border. Robert Albro, a researcher at the Center for Latin American and Latino Studies at American University says, "The main reason people are moving is because they don't have anything to eat. This has a strong link to climate change—we are seeing tremendous climate instability that is radically changing food security in the region."[27] In June 2019, *The Guardian* profiled the story of Honduran Jesús Canan, who described how an extreme drought "is forcing us to emigrate. In past years, it rained on time. My plants produced, but there's no longer any pattern [to the weather]."[28]

"There is no longer any pattern" is not just a description of the weather but of the vacuous state of contemporary politics that will have to come to grips with unprecedented waves of extreme weather and mass migrations in the near future. Even the fairly conservative World Bank estimates that warming temperatures and extreme weather will force 3.9 million migrants to flee Central America over the next thirty years alone, and an estimated 150 million to 300 million climate refugees are set to be displaced worldwide by 2050.[29] The IPCC predicts that at 3.6 degrees of warming, there will be a "'disproportionately rapid evacuation' of people from the tropics. As Aromar Revi, director of the Indian Institute for Human Settlements reports: 'In some parts of the world, national borders will become irrelevant. . . . You can set up a wall to try to contain 10,000 and 20,000 and one million people, but not 10 million.'"[30] While countries such as Bangladesh are most vulnerable to these mass displacements of people, it isn't difficult to foresee a near-term future of the American Southwest that will potentially have to grapple with millions of migrants while also "adapting" to climate breakdown.

The current wave of migration from Central America and the development of the CAAP were co-occurrences in the same space/time, a literal passing one another in the streets. Yet, for the most part, the CAAP ignores the impact of refugees coming to San Antonio in the next fifty years. The CAAP says San Antonio should "periodically review the City's ability to provide for the needs of coastal hurricane evacuees and other populations displaced by extreme weather and climate events."[31] But how many displaced people should cities like San Antonio expect to be able to provide for? As an inland city with close proximity to Houston and New Orleans, it will likely be a first choice for residents who decided to flee from these cities. This was the case when Hurricane Katrina hit in 2005, and San Antonio took on an estimated 30,000 evacuees. It was the second-largest resettlement after Hurricane Katrina, and many settled in San Antonio neighborhoods permanently.

While fewer Houstonians relocated to San Antonio after Hurricane Harvey, Houston is the fourth-largest city in the United States and has a long history of hurricanes. In the next generation, San Antonio's major concern might not just be heat waves, droughts, and flash floods, but a wave of migrants from low-lying coastal cities like Houston (fifty feet above sea level) and New Orleans (twenty feet above sea level), and waves of migrants fleeing climatic and political crisis in Central America. At present, the United States handles this problem by treating immigrants as criminals, separating children from their families, and placing them into cages in detention centers, many times for private profit. There is no better example of contemporary social orders based on coloniality. But what happens when those migration numbers increase one-hundred-fold? If at present, cities like Houston cannot even come close to adequately responding to Hurricane Harvey—many residents are still waiting for their homes to be bought out or repaired—then what will happen when things get much worse in the coming decades? While foreign migration and domestic resettlement may seem like isolated occurrences now, one can easily imagine them occurring simultaneously. The macroversion of this story points to the limits of the liberal democratic project to absorb these populations. At this scale, it's too easy to see how containment too quickly becomes something akin to the worst atrocities in human history.

The question of migration is not just for the dispossessed but also for those with the most privilege who continue to believe that one can adapt to climatic changes by moving farther north. Such privileges are alive and well when major popular science journals run articles that advertise the twenty places you should move to or invest in to survive climate change.[32] Such practices reinforce the structural logics of inequality that will lead democratic civil society to versions of the hunger games that pits tribes against tribes and families against families in an ultimate battle for survival that the elites watch on screens from their rebranded equity towers. And in the world of the hunger games, if you think you can survive alone with your family on twenty acres, you may want to watch again.

A dystopic future of climate adaptation wherein entrenched privileges breed even more radical forms of segregation and ignorance to the experiences of marginalized people is a fast-approaching reality of climate apartheid. This form of adaptation can take dramatic turns toward the kinds of assumed privileges that solidify nationalism, erode global organizations formed after World War II, further criminalize migrants, and jail activists for protesting against extractivism. Currently, US national politics are awash with republican denial over climate change despite overwhelming evidence. In rhetorical circles, too much ink has been spilled over framing this issue as an issue of science when it's not.[33] Rather, denial, like the aversive forms of racism from the Aquifer Protection Association in chapter 3, is a form of protection and self-care for communities with vested interests in current sociopolitical arrangements.[34] As this denial continues to break down, intellectuals continually consider a dystopic future wherein climate adaptation becomes a version of sociopolitical reform in the worse possible ways—a reanimation of bordered nationalism built on twenty generations of coloniality, slavery, and genocide, and the forms of apathy and schadenfreude perversions that allow these practices to continue. You won't find these concerns mapped in psychometric climate change polls. But as it becomes even more clear that climate adaptation will mean mitigation of emissions on industry and business, adaptation to new forms of extreme weather, and the technological capacity to democratize renewable energy, it is the legitimization and validation of the public practices of shared sacrifice, public service equity, place-keeping, and inhabitation that will help keep this dystopic future at bay.

As all of us live through the fifth great extinction on earth, humans will certainly lose many forms of symbiosis (hopefully not the obligatory ones like cereal grains) that potentially place human survivability at stake. There's a devastating circular logic at work here wherein the processes of individuation to extract and accumulate eventually are self-defeating—the logic of a cancer metastasized by climate breakdown. Already the military of the United States is fully considering the cascading effects of millions more migrants all living through extreme weather, all demanding access to basic services like clean water, energy, and food. Already the US military is considering containment strategies not just for mass migrations of people but for the inevitable outbreaks of infectious diseases.[35] Migrants, refugees, and asylum seekers; the experience of multiple extreme weather events simultaneously; the demand placed on fragile public service infrastructures; and the spread of infectious diseases across populations of people will overwhelm rhetorical theory and practice with its exponential happenings. Digital technologies will not only be necessary, they will also function as technologies of surveillance and neocolonization of displaced peoples, as the work of Ruben Casas shows.[36] The intersections of decolonial rhetorical science studies and immigration rhetorics have a way to go, but the scholarship of Karma Chávez, Ruben Casas, and more will be central. For rhetorical science studies, the potential in countering these neocolonizations must be done with solidarity. And the only way to build solidarity is to engage in practices with migrants, refugees, and asylum seekers and use the tools of research, teaching, and service to assist them. Rhetorical futures of migrations can never be about climate effects alone, but rather how current public, political, and personal narratives interact with the situated technoscientific knowledges of digital media, human/environmental health, and climate instability. Future cases of migrants may very well show that the mass mobilization of human bodies demanding basic services on an unstable planet may be an inevitable fate. When the tap water stops running, when the electricity is shut off, when there's a run on food and gas, and the extreme weather event hits again, we will all be migrants.

ENGAGED RHETORICAL CRITICISM IN ENERGY MICROPOLITICS

If the auguries of climate science are anywhere near correct, the levels of violence unleashed by climate breakdown will be unprecedented in their scale, speed, and severity. It's still not exactly clear how any public entity is supposed to face up to and grapple with these technical facts. As a document that attempts to do this, the CAAP lists a variety of lofty adaptation strategies that seek to build capacity for infrastructure resilience, public health systems, emergency management/community planning, food security, just energy transitions, and of course, protecting property for progressive economic growth. In a section titled, "Why Adapt?," the CAAP asserts, "Without adaptation actions to improve our resilience to these impacts, 'climate change is expected to cause growing losses to American infrastructure and property and impede the rate of economic growth over this century.'" And "the vision for San Antonio's future is that of a resilient city, meaning a city that can maintain normal function in response to external stresses and disruptions, specifically those from climate change."[37] In this framework, adaptation builds capacity for social services, but it does this primarily to provide protections from economic loss. For the CAAP, climate change is less a cause for resilient thinking about sociopolitical transitions than a recalcitrance back to a preconceived "normal."

But if cities truly listen to climate science, they might understand we are at the end of normal, and the question of human response-ability is a question of acknowledging and acting on these changes. For city-level plans like the CAAP to equate resilience to a maintenance of normal function underestimates the systemic cascades of disruptions climate breakdown will induce. Normal feels too much like victory, when victory might be beside the point as cities grapple with global forces beyond their control. Too often city plans like the CAAP cannot face up to the limits of local politics, which are only ever strong enough to shape national and global forces, never fully resist them. As this book has shown, global forces are constitutive of what we call local, and yet the power of the local is to fold these forces of globalization in ways that create some measure of shared prosperity. Rather than a victory to return to a normal state, the chapters in this book have found that true victory might be achieved through a "multiplicity of miniscule acts of care and attention."[38] The structural view from climate science is very real, but something always escapes. Nature finds a way and so do people. Local climate politics must attend to the quiet ways in which individuals and groups define living well together at the end of normal.

Living well at the end of normal must be registered through rhetorical theory and practice, too. Rhetorical projects that call for more rhetoric and more deliberation at the scale of the local, for example, fail to recognize the real lesson behind democratic participations' lack—willingly or not, people must trust institutions to make consequential decisions for them. The fundamental lack of participation in democratic politics cannot be addressed through calls for more rhetoric. Part of the reason why populist rhetoric is so persuasive throughout history is because it simplifies the complexities of politics through a strong charismatic leader who people believe will make decisions that make their lives easier. Most working people do not want perpetual civic engagement; rather, they want some anonymous power to make decisions for them so they can watch Netflix and write books.[39] The agreement of the social contract, after all, based as it is on the political philosophies of John Locke and Jean-Jacques Rousseau, means that subjects agree to be ruled and dominated in exchange for protections and privileges like access to clean water, reasonably priced power sources, decent transportation options, basic security, and a meaningful say in the decisions of the city-state. If working people vote at all, they do so in good faith that they won't have to think too hard about basic public services and securities that help foster the good life. The lack of participation in local democratic politics feeds on the notion that not everyone can engage in the constant work of tending their own garden, when they are already too tired trying to get food on the table. Technocratic capital's ability to move the socioeconomic levers to just barely below full revolt from the working and middle class, while simultaneously engaging in misinformation campaigns that transfer the blame to government itself, drives the lack of participation in democratic processes. What can rhetorical theory and practice achieve in this political climate?

Rhetorical studies of everyday and emplaced politics have shown how at the core of democratic practice is an agonism and an anarchism wherein appeals to social justice, to the marginal, to rights, while powerful, point to a flexible moral core of democracy that can be deployed to support any material condition and ideological position.[40] But this double-edged blade of democratic rhetoric should not leave politics ambivalent. The ratios of democracy matter. The point of public rhetorics and local politics is to move those ratios toward the best and broadest possible version of equity and justice for the here and now. Rather than appealing to a transcendent democracy that is inherently moral, good, and just, scholars of micropolitics have suggested that political agency is found in small actions that make a difference, and that there is no justice except for the kind one helps to create. Addressing the structural inequities in one's own backyard doesn't leave one ambivalent but rather resolute in working toward matters of arrangement for distributive, representational, and procedural justice in affordable housing, flood projection, fair energy rates, high-quality and efficient transportation, and more. These are the structures that can help foster more of those miniscule yet meaningful acts of care and attention.

Similarly, the rhetorical turn toward micropolitics points to the limits of those who continue to make grand claims about decoloniality in favor of the everyday work of struggling through colonial structures. As decolonial rhetorical scholars continue to call for pluriversality as a global project, they would do well to embrace the complexities of micropolitics, since they are also never without others. As Candice Rai points out, moments of real social and political progress—the emancipation proclamation, the civil rights act, the environmental protection act—are never more than "a stable-for-now social state that might just as easily backslide as improve in the future, and, moreover, [that state] is always contested and . . . always benefits people unevenly across the social landscape."[41] Decolonial rhetorical projects require a complex engagement with the concrete pragmatics of micropolitics rather than the all too common simplicities of ethno-identity politics that play out dangerously across ethnic groups. This book has examined moments in San Antonio's history in which climate has played a significant material-semiotic role in shaping the city's politics. The events in this book make clear that the long-standing fight for sociopolitical equity from Mexican American communities have led to a progressively higher quality of life for working-class families. Each chapter in this book has pointed to ways in which politics could have been directed much more in the spirit of decoloniality—equitable flood protection, nonaversively racist coalitions, a full embrace of Chicana/o aesthetics, a pragmatic commitment to equity at the expense of capitalistic development, and so on. Still the gains toward racial parity and inclusion at the level of the social and the political are real, even if they are incomplete.

This book has tracked the climate politics of San Antonio just through one version of its extreme climate—its flooding—and arguably it is the heat waves and droughts that may prove just as damaging to this city's ability to "adapt." In the decades ahead, San Antonio will certainly experience a few more weeks of 90+ degree temperatures and will begin to see the emergence of 110+ degree days and 80+ degree nights, which may be standard in arid Phoenix, but they are rare in the humidity of a southern-like summer in San Antonio. Additionally, consider the urban heat island effect, which can mean around a 20-degree Fahrenheit difference from the central city to the rural reaches of the city. Urban heat islands are another example of how the most vulnerable of residents—the poor, elderly, working-class—who can least afford to adapt to a crisis they barely contributed to, will also be most vulnerable to its effects. In these conditions, the six stages of hyperthermia—heat stress, heat fatigue, heat syncope, heat cramps, heat exhaustion, and heat stroke—will become a too common experience for working and poor people. In the same way Quiroga captured a sense of the colonial landscape as violently anti-Mexican, the local effects of climate change will mean monstrous weather patterns that reanimate colonial processes of domination against working people even at the level of physiology. But climate change in Texas is not just more heat waves. Future adaptation strategies will have to address an intensely paradoxical climate with more intense drought, precipitation, and polar vortex events that threaten to disrupt every aspect of the modern/colonial world.[42] And in a world where climate is constitutive, rhetorical action will bind climate anxieties.

It's already clear that local action around the micropolitics of distributed energy systems must be a major force of work ahead given the wicked combination of increasing extreme weather events and a fragile national power grid. Fragile power plants, transmission and distribution systems, and a lack of significant investment in the power grid mean communities will have to start thinking seriously about decentered forms of power generation and distribution, especially with increased energy requirements due to extended periods of heat, drought, and cold. As the case of the October/November 2019 California fires demonstrates when PG&E cut off power to more than one million residents, and as the case of the February 2021 Texas polar vortex demonstrates when ERCOT cut off power to millions of residents, some of whom died from freezing temperatures, indiscriminate power shutoffs will be more common in order to protect people from extreme hazards. Local pictures can be grim, but as a recent report by the US Army notes, any serious collapse of the national power grid would be catastrophic with significant national losses in food, medicine, water/waste distributions, communication, transport, and fuel systems, and more.[43]

Given these realities of climate breakdown, pluriversal energy rhetorics and politics will likely be a major focus of work moving forward. The rhetorical scholars who have anticipated this work have written a fine review to suggest that energy must be a focus of attention not just during times of crisis but also in everyday life.[44] Traditionally, rhetoricians who study energy have done so only after a crisis event like Three Mile Island or Fukushima, but Danielle Endres and her colleagues suggest an examination of the internal rhetoric of energy sciences, comparative studies across energy type, and energy in everyday life.[45] Citing a special issue of *Theory, Culture, and Society*, the authors note that everyday practices around energy are starting points for just and sustainable transitions. Rather than fall into resilience traps that support the status quo through appeals to coal's reliability and stability, the authors suggest that micropolitics would allow one to minimize demand for energy while at the same time building neighborhood-level capacities to generate and store energy so these systems are not so reliant on expensive and fragile national grids. After all, what would make one consider everyday energy more than a daily check of a storage battery, or driving by the local neighborhood energy hub? Perhaps smart homes might be those that know when to turn most of the energy off rather than perpetually burning the candles. Understanding these micropolitics of energy at the level of constitutive material-semiotics seems essential. As the paradoxes of local and global are worked out through rhetorical performance, rhetoric should be attuned to the pluriversal potentials of decentralized systems that learn from heterogenous worlds of energy that also negotiate their potential articulation when needed.

EMBRACING DISTURBANCE: PLURIVERSAL RHETORICS AFTER PROTECTION AND PROGRESS

It's dramatically apparent that flooding in South Texas will increase in both severity and intensity over the next twenty-five to fifty years at a time when urban growth patterns will also place more people at risk. San Antonio is expected to grow by one million people in the next twenty years, and urban development along the one-hundred-year floodplain is still common, even though that figure is no longer a sufficient land-use planning guide. The National Oceanic and Atmospheric Administration (NOAA) recently updated its historical flood records through its Atlas 14 project, and it reports that "historically calculated 100-year flood models are now occurring closer to once every 25 years, meaning such flooding is four times more likely to occur than previously predicted."[46] NOAA's Atlas 14 data will be used by FEMA for their own flood insurance rate maps, yet the politics behind such maps will likely mean distortion in order to prop up the market value of homes, artificially if they have to. Meanwhile, the national rate for flood insurance in the United States is the lowest it has ever been. What will this mean for the more than one million people who will live in or even near floodplains? How will cities like San Antonio build for protection with any measure of equity to keep stories like La Tragedia from happening over and over again?

One of the calls to action that appears early on in the CAAP is to act on extreme weather threats to the insurance market. The CAAP cites a 2019 survey of 247 insurance actuaries who identified climate change as the top emerging risk—higher than cyber threats, financial instability, and terrorism. In its adaptation strategies, the CAAP focuses on early warning systems, flood awareness, and creating "flood-proof" critical infrastructure. They say San Antonio will "work with responsible parties to identify and undertake prioritized retrofit programs for critical infrastructure (transportation, building, IT and telecoms, utilities sectors) to ensure resilience to flood impacts over the lifetime of the asset, once updated floodplains are available and also incorporating future climate projections."[47] This is not an unambitious undertaking as many transportation corridors, buildings, and utilities already exist within the one-hundred-year floodplain. So, how is "flood proofing" supposed to work when the new norm is a one-hundred-year event every quarter century? If there were any indicators of how the "flood-proof" adaptation strategy is going, it might be the national flood insurance market itself. As the *New York Times* has reported, the National Flood Insurance Program, which is run by the Federal Emergency Management Association (FEMA), covers about 95 percent of all residential flood policies in the United States. But since the early 2010s when Congress allowed the agency to let rates rise to more closely reflect the true risk of flooding, the number of insured Americans has steadily declined. In some midwestern states at the greatest risk of flooding, the numbers of the insured are below 15 percent. Nationwide, only about a third of all homes built in the floodplain have insurance, despite an agency goal of doubling the number of flood policies by 2023.[48] While there's a difference between home insurance and critical infrastructure, reading the CAAP's adaptation strategies in light of the National Flood Insurance Market leaves one wondering that if the federal government cannot adequately protect and insure homes (assets) and people, how exactly is a city government supposed to ensure/insure the protection of critical infrastructure?

The CAAP falls silent on the option to buy out homes and neighborhoods built in the floodplain, but it's clear that what's coming for those who live in or near floodplains is a series of difficult choices—which homes and neighborhoods should be protected and which ones should be bought out? Which historic structures should we save and which should we let go of? Jeff Goodell argues that "smart cities will develop master plans, articulate long-term strategic visions, revise zoning ordinances, pass tax incentives to shift development to higher ground, . . . and that's just a start."[49] In some cases, retreat will be the right answer. And retreat requires so much more than a techno-fix like geo-engineering. It requires foresight, individual action, and a willingness to change your life. In response to Hurricane Harvey, Houston instituted a new development standard that requires rebuilding above the five-hundred-year floodplain. Projects like the thirteen-mile-long San Antonio River linear park and the two-mile-long San Pedro Creek Cultural Park are both good examples of using flooding as a cause for thinking about urban design so cities can "slow it down, spread it out, and let it soak in."[50] Projects like the San Pedro Creek Project are a good example of what adaptation looks like today—an expensive restoration of wetland and riparian areas as green spaces to collect rainwater, plant pollinator gardens, and practice place-keeping. But not all areas of town can be treated with the same level of attention, care, and investment. Inevitably, some areas of town will be left to fend for themselves when the next hundred-year event or worse hits.

A truly transformative pluriversal adaptation would mean cities begin to develop whole new ways of relating to flash floods in order to foster resilience as shared sacrifice with catastrophic floods in a climate-changed world. At the ground level, this means buying, relocating, and/or tearing down buildings and homes in flood-prone areas (after experiencing two or three floods every decade, most people are willing to consider a buyout), and passing and enforcing growth policies on where and how buildings and homes get built. Many factors will go into the decisions about what or what not to protect, but the larger point of shared sacrifice is an acceptance that not everyone, not every neighborhood, is going to be saved. Attending to those displaced with any measure of equity will be a major challenge. When the 1921 flood leveled San Antonio, the mayor was adamant that it would not accept outside money for help; the future iteration of this history is the day when people and neighborhoods realize that their local, state, and federal government does not have the money or political will to rescue them, and they'll be left to fend for themselves. In some cases, the human choices around displacement will come from publics, not necessarily politicians. For politicians, there is an incentive to encourage staying in place—it's personally difficult to abandon a home, it's time-consuming, it's politically messy, and it costs money, not least by eroding a tax base. After a disaster hits, there are many incentives to rebuild, but few incentives to build differently. Building codes, zoning regulations, FEMA floodplain designations, and more all incentivize a static and domineering relationship to land and water that assumes protection. At the very least, the policies in place that drive development will have to learn to live with flooding, even in potentially productive ways, rather than assume protection. To learn to love these floods would mean that a culebra de agua is not monstrous, but rather a kind of embodiment, identity, and life that fosters repatriation of land back to nature, and back to Indigenous communities and ways of life.

The broader point is that rhetorical work on floods, and any other extreme weather event, at the end of insurance will not be an art of progress based on protection. It will, rather, be a diverging art, a pluriversal art of new meaning-making with floods in each location. This is to say that public rhetorics through extreme weather events will likely be characterized as a pluriversal art where every location must make meanings anew. And every location will come to know what floods are saying to them, as more frequent extreme weather events perform with enough rhetorical force for humans to keep floods in mind at all levels of culture, community, and city-level policy. These are yet more reasons why microclimate politics are ever more important. As every location looks into the past and futures of extreme weather— fire, floods, heat, sea-level rise, and such—all cities will have to consider how to live with and love these events rather than resist them as disasters. In other words, climate will carry rhetorical force because it maintains a relationship to materiality through everyday practices, institutions, and to public spaces where bodies, genres, and vernacular discourses carry it, circulate it, maintain it, and reproduce it, in order to keep it in mind.[51]

Loving the monsters of extreme weather events at a time when it is apparent that government is moving too slowly to adequately protect its people will likely mean an explosion of weather ways in relation to extreme weather events that will never be without others, or without the digital supercomputers in our pockets. Such rhetorical work of negotiating life in common through climate breakdown might consider rethinking notions of disaster as disturbances along the lines of ecological theory. As I mentioned in the preface, such work might leverage Anna Tsing's work on disturbance theory where she reminds us that "no single standard for assessing disturbance is possible; disturbance matters in relation to how we live. . . . Disturbance is never a matter of 'yes' or 'no'; disturbance refers to an open-ended range of unsettling phenomena."[52] Words express relations to the world, and unlike the word *disaster*, disturbance is not just characterized by loss, or the inability to cope. Though unsettling, disturbance is also profoundly creative as causes for thinking about relations to place and environment. Disturbances may cause blackouts, but they also are a cause for relational thinking about energy. Disturbances may mean the end of normal, but they also mean an opening up of possible worlds among worlds in relation to extreme weather events and their effects. Disturbances are pluriversal and rhetorical in nature, whereas disaster is associated with irrevocable loss from modern/colonial patterns of living. The roots of disturbance mean an interference of normal arrangements or functioning, which would seem to have some positive valence for de/coloniality as an analytic and a practice. Some species are more adapted to disturbance than others, and some are more used to disturbance than others, though they may undergo transformations during the process. Perhaps this is another augury of shared practices that address long-standing histories of colonial relations to land and people through pluriversal rhetorical praxis: the vulnerabilities and creative capacities of loving the monsters of disturbance.

Coda

Pluriversal Rhetorical Projects

I SUBMITTED A FULL VERSION of this book project in January 2020. When I went back to it after the peer review process, it was the end of April 2020 and COVID-19 had disrupted all of our lives. As the Pentagon report I cited in the last chapter had predicted, a major outbreak of an infectious disease blossoming into a pandemic was not just predicted as possible but inevitable. Based on similar reports, the last chapter discussed a few other foreseeable futures, some of which have been called inevitable. As I submit this book for the last time in September 2020, with more unprecedented fires raging across the West Coast and more slow-moving torrential hurricanes landing throughout the Gulf of Mexico, I can't help but wonder what the world will look like a year from now, ten years from now. Potentially, COVID is a once in every hundred-year event. Then again, potentially, these kinds of disruptions are just the beginning of cascades of violence on a thoroughly damaged and increasingly uninhabitable planet. People, listen to what your earth is saying!

What are human response-abilities to these mostly foreseen, completely self-induced, and increasingly dire conditions? The facts are in. The world knows what to do. The world knows how to act. On these matters, the science is clear, but the shared practice is not. This is partly why the Black Lives Matter movement is not just about civil rights and racial justice any longer. All social movements now are a response to the systemic injustices of coloniality/modernity that have pushed everyday people to the brink. For rhetorical studies, I have argued for the need for pluriversal futures. And as I mentioned in the introduction, pluriversal rhetorics attempt to foster worlds among worlds characterized by divergence, heterogeneity, interdependence, and co-constitution. The only way to truly accomplish this is an attempt to break down oppressive structures of everyday coloniality. Doing this is what would allow for the radical notion that anyone and everyone can contribute to decoloniality as long as it works against everyday coloniality and builds space for knowledges and practices of the historically oppressed: the Indigenous, Black, Chicanx, and many other peoples of color. It is difficult to do this in a monograph (an institutional requirement for many professors trying to survive), but not impossible. Whether I've succeeded or failed on this point is a question I'll leave to my critics, but in this coda I want to describe a few pluriversal rhetorical projects underway that I believe will be important for the future of rhetorical studies writ large.

Though it wasn't billed as such, the first Rhetoric Society of America Project on Power, Place, and Publics is a fine example of how local sites foster pluriversal inquiries across Indigenous publics (Haas), political economy (Greene), disability (Vidali), community-based roots (Royster), environmental justice (McGreavy), and so many more approaches that touch on the everyday legacies of coloniality. The grant-based work that rhetorical ecologists engage in can play a vital role in doing this critical pluriversal work. For example, Bridie McGreavy's project with Darren Ranco and the Penobscot Nation in Maine examines ecological restoration in association with land return and Indigenous sovereignty. Such work demonstrates how pluriversal projects can work to meet the most difficult criteria of decoloniality as a political project for repatriation. A related example is the work of Caroline Druschke and Kassia Shaw, which examines a co-managed prairie site in northern Wisconsin that was returned to the Ho-Chunk after its use as an ammunition plant. By working directly with the Ho-Chunk, Kassia Shaw hopes to examine dimensions of pluriversal projects that combine rhetoric and science with the traditional ecological knowledge of the Ho-Chunk. Of course, pluriversal projects will be led by Indigenous and other scholars of color in rhetorical studies, perhaps in some cases exclusively so. But these projects are good examples of how white settler scholars in rhetorical studies can also contribute to creating space for pluriversality and decolonial politics inside and outside rhetorical studies.

Though not all pluriversal rhetorical projects will achieve repatriation, this criterion is not the only way such work can create space for Indigenous and other ethnically specific knowledges, practices, and communities. One of the outcomes of this book project, for example, is a grant-funded pluriversal project that combines avian ecology and rhetorical science studies with Mexican American studies and Indigenous studies. By using backyard bird feeding and rhetorical approaches to community-based science, this aspect of the project hopes to teach students and community members alike how to engage in digitally augmented observation, fieldwork, and experimentation in order to learn about scientific methodologies and their increasing need for community-based participation.[1] At the same time, the project offers students and community members' knowledges and practices from elsewhere and otherwise. Our Mexican American studies program is led and guided by community-based scholar Dr. Marisol Cortez, whose scholarship, creative writing, activism, and subsequent workshops will engage participants in the histories and futures of Chicanx civil rights and politics in San Antonio and beyond. Our Indigenous studies program is led and guided by Dr. Annette Portillo, whose scholarship and subsequent workshops engage Indigenous women's storytelling practices in relation to place and landscape. The project also consults with Coahuiltecan elder and member of the Texas Native American Church, Masauki Celso Zepeda. He, along with his partner, Lupita Lugones, director of the Mariposa Centro Holistico in Toluca, Mexico, will run workshops on the Indigenous knowledges of birds in South Texas and Northern Mexico. For each workshop, Masauki and Lupita will perform sacred Indigenous rituals with drums, banded feathers, and dancing signifying the relationship Coahuiltecan peoples have cultivated with the birds of this land for thousands of years. They will engage community participants in ceremonies that cultivate an interdependent relationship with birds, increase knowledge of land engagement, and promote conservation behaviors. Lupita's work in Mexico has given her access to the avian conservation work of puppeteer Heather Henson, daughter of puppeteer Jim Henson. Heather Henson's work focuses on crane conservation, and given that cranes have a long and storied history in South Texas, these workshops will likely include dramatist displays of crane puppets, which may also be used in the final student symposium. While not achieving land repatriation (in Texas!), the team believes that creating space in these ways can build momentum for sciences from below and decolonial politics.

Scholars and writers outside of rhetorical studies have been doing this work for many years, and it would benefit rhetorical studies to learn from them. Listen to Robin Wall Kimmerer and how she describes scientific research as a ceremonial act:

> Because we can't speak the same language [as our more-than-human others], our work as scientists is to piece together the story as best we can. We can't ask the salmon directly what they need, so we ask them with experiments and listen carefully to their answers. We stay up half the night at the microscope looking at annual rings in fish ear bones in order to know how the fish react to water temperature. So we can fix it. We run experiments on the effects of salinity on the growth of invasive grasses. So we can fix it. We measure and record and analyze in ways that might seem lifeless but to us

are the conduits to understanding the inscrutable lives of species not our own. Doing science with awe and humility is a powerful act of reciprocity with the more-than-human world.[2]

Listen to Gloria Anzaldúa's teachings about how to shift toward a path of conocimiento . . .

> With awe and wonder you look around, recognizing the preciousness of the earth, the sanctity of every human being—somos todos un paíz. Love swells in your chest and shoots out of your heart chakra, linking you to everyone/everything—the aboriginal in Australia, the crow in the forest, the vast Pacific Ocean. You share a category of identity wider than any social position or racial label. This conocimiento motivates you to work actively to see that no harm comes to people, animals, ocean—to take up spiritual activism and the work of healing. Te entregas a tu promesa to help your various cultures create new paradigms, new narratives. . . . The life you thought inevitable, unalterable, and fixed in some foundational reality is smoke, a mental construction, fabrication. So, you reason, if it's all made up, you can compose it anew and differently.[3]

Pluriversal rhetorical arts can learn from this composition done differently. And it can do so by reckoning with coloniality, taking up the work of healing, and fostering many sciences, many rhetorics, and many worlds.

Acknowledgments

MY DEEP APPRECIATION FOR YANAGUANA/SAN ANTONIO comes first. I arrived in San Antonio in May 2015, and our first walk was from the Beacon Hill neighborhood to San Pedro Springs Park—one of the oldest parks in the country and long a gathering place for all of San Antonio's inhabitants. I acknowledge all of those who have practiced with and protected these places, from the Payaya and other Coahuiltecan tribes, to the Lipan Apache and Comanche, and to the Spanish, Texians, Americans, and everyone who cares deeply and acts responsibly to protect public spaces. I had never dwelled in a place where the headwaters were located so deep in the heart of a city, and for me, this is the primary metaphor for beloved San Anto—a magical source of spring-fed water as a center for all the city's inhabitants. Through this book, I have attempted to acknowledge legacies of colonial violence that all of us settlers are implicated in, myself included, but I have also attempted to put into equality the worlds among worlds that compose the cosmos of San Antonio. This book and the projects that come from it pledge to work against one-world extractivism and to work for pluriversal futures, now and forever.

I have wanted to write a nature-culture book ever since I can remember, but especially since my exposure to the Nature and Culture program at the University of California, Davis, during my undergraduate years. This environmental studies program was started by Gary Snyder and a handful of scientists, social scientists, and liberal arts scholars/professors who designed the education they would have wanted for themselves. Classes were co-taught by science, anthropology, and English professors, and more; we engaged in ecological fieldwork projects; they required Indigenous literature courses; internships were commonplace; and we had just enough scientific training that many students went on to advanced degrees in ecology. Though the program no longer exists, it's clear this curriculum was well ahead of its time, and most of my academic and community-based work has attempted to re-create this experience at my own institutions. I am deeply grateful to those scholars and teachers who offered me this early transdisciplinary training—Scott McLean, Shorty (Virginia) Boucher, Michael Smith,

Michael Barbour, Peter Moyle, all my peers, and so many more. No program is perfect, but Nature and Culture was a wonderful early example of how to build a pluriversal curriculum to meet the challenges of climatic and environmental response-abilities.

To my mentors, friends, and co-laborers across rhetorical studies, I would not be here without you—Lynda Olman, Scott Graham, Bridie McGreavy, Caroline Druschke, Greg Wilson, Ashley Rose Mehlenbacher, Leah Ceccarelli, Damian Pfister, Jennifer Malkowski, Jean Goodwin, Carolyn Miller, James Wynn, Carl Herndl, Ken McAllister, Amy Kimme Hea, Lauren Cagle, Damián Baca, Anne-Marie Hall, Cruz Medina, Amanda Fields, Cristina Ramírez, Lucía Dura, Aimee Roundtree, Jane Detweiler, Crystal Colombini, Ruben Casas, Ralph Cintrón, Candice Rai, Christa Olson, Derek Ross, Kyle Boggs, Phaedra Pezzullo, and so many more. All of you have been an influence across my academic journey, and I hope you can find some aspect of your work validated in these pages. Thank you.

I was hired into a department that was deeply implicated in the politics of white privilege, and it took me a few years to figure out how to navigate it. To my students, colleagues, and friends who helped me see this and fight against it, I could not have survived this battle without you, and I am deeply appreciative of your support and solidarity: Jasmine Hale, Sonja Lanehart, Kinitra Brooks, David Vance, Ben Olguín, Gabriel Aguilar, Annette Portillo, Kerry Sinanan, Kimberly Fonzo, Sonia Saldívar Hull, Jackie Cuevas, Kim Garza, Joycelyn Moody, Michael Gallaway, and others. I am also eternally grateful to my colleagues and friends in UTSA's Department of Environmental Science and Ecology and our broader grant team, all of whom are invested into creating equitable and inclusive institutional structures: Jennifer Smith, Claudia García Louis, Janis Bush, Amaury Nora, Eres Gomez, Amanda Lamberson, Sarah Gorton, Jamie Killian, Marina Zannino, and more. You all have made my time at UTSA so much more enjoyable than this setup of a job would have been otherwise. Thank you all for your support, love, friendship, and solidarity. Onward!

To my San Antonio friends and family—you have contributed so much to this project whether you know it or not and I thank you: Marisol Cortez, Greg Harmon, Jessica O. Guerrero, John Phillip Santos, Tomás Ybarra Frausto, María Antonietta Berriozábal, Ana Sandoval, Dianne Moffett, Marissa Ramirez, the entire Hernandez family, Ed Conroy, Masauki Celso Zepeda and Lupita Lugones, Los Nahualatos, and more. Thank you for making my life more complete and for the convivios over the years. Without you and your work, this project would not exist. It is an honor to know you and appreciate your gifts to the world.

Chapter 3 was previously published as "Divergence and Diplomacy as a Pluriversal Rhetorical Praxis of Coalitional Politics" in *Rhetoric Society Quarterly* 50, no. 4 (2020): 225–39. I'd like to thank the College of Liberal and Fine Arts at UTSA for a subvention grant that supported this book project, and to the Department of English at UTSA for the course releases required to finish this work. This material support has been essential.

Finally, to my family, Mary and Steve Walker, Nate Walker and Vikram Paralkar, Soozie and Marty King-Kostelac, Damien King-Kostelac and Valerie Wingfield, my cousins, aunts, uncles, grandparents, and all my kin. Most especially to Amelia King-Kostelac, Theodore Walker, and Noura Walker—my lights, my loves, my everyday acts of carefulness through difficulties and beyond, I will forever count the beats with you.

Notes

PREFACE

1. Santos, *De Unos Lugares Perdidos*.

2. De/colonial is an attempt to reckon with coloniality and acknowledge decolonial possibilities. I use it here to reckon with complicity in colonial practices that connects and differentially affects everyone while also acknowledging alternatives. Wanzer-Serrano, *New York Young Lords*. Climate breakdown is a term George Monbiot coined in the wake of the IPCC's release of AR4 to describe how the climate variability, extreme weather events, and general climate conditions that fostered human/nonhuman evolution on this planet are now collapsing. Monbiot, "Climate Change? Try Catastrophic Climate Breakdown," last modified September 27, 2013, *The Guardian*, www.theguardian.com.

3. Miller, *San Antonio: A Tricentennial History*, 27.

4. Miller, *San Antonio: A Tricentennial History*, 28.

5. Barr, "Beyond Their Control," 150.

6. Haggard, "Spain's Indian Policy in Texas," 76–77.

7. Schneider and Panich, "Native Agency at the Margins of Empire."

8. Harding, "One Planet, Many Sciences," 39–62.

9. Miller, *San Antonio: A Tricentennial History*, 2.

10. Miller, *San Antonio: A Tricentennial History*, 6.

11. John Phillip Santos, personal correspondence, May 2019.

12. Santos, "San Antonio Is a City of Metamorphosis."

13. Tsing, *Mushroom at the End of the World*, 161.

14. Gibbons, "NOAA Study Shows Heavy Rains Have Grown More Frequent in Texas."

15. Sharif, "Climate Projections for the City of San Antonio."

16. Fourth National Climate Assessment, vols. 1 and 2, November 2018, www.global-change.gov/nca4.

17. Intergovernmental Panel on Climate Change (IPCC), "Summary for Policy Makers."

18. Wallace-Wells, *Uninhabitable Earth*, 11–13.

19. Martinez, *Injustice Never Leaves You*.

20. Pezzullo, *Toxic Tourism*, 20.

21. Olson, *Constitutive Visions*.

22. Alcoff, "The Problem of Speaking for Others," 24–27.

23. Olson, *Constitutive Visions*, xvi.

INTRODUCTION

1. Najda Popovitch, John Schwartz, and Tatiana Schlossberg, "How Americans Think about Climate Change, in Six Maps," last modified March 3, 2017, *New York Times*, www.nytimes.com. Yale's 2017 US Climate Public Opinion Map lists 47 percent for Bexar County, 46 percent for Marin County.

2. Housing Rights and Climate Justice, *Climate Action SA*.

3. Housing Rights and Climate Justice, *Climate Action SA*.

4. Anzaldúa, *Borderlands/La Frontera*.

5. Housing Rights and Climate Justice, *Climate Action SA*.

6. "Supercharged" is the word climate scientist Michael Mann uses to describe extreme weather events that are made more intense by human-induced climate change.

7. Montejano, *Anglos and Mexicans in the Making of Texas*, 40. Historian David Montejano points out that the first city council of San Antonio was entirely Spanish surnamed, but by 1850 the council was almost completely replaced by German, French, and English surnames. The fate of Mexicanos during this period is exemplified by the story of Juan Seguín, San Antonio's first Mexicano mayor in 1841. There would not be another until Henry Cisneros in 1981.

8. "Hot Summers Long a Way of Life," *San Antonio Express News*, August 28, 2018, 13.

9. "Atlas 14, Volume 11," National Oceanic and Atmospheric Administration.

10. Lind, "'500-Year' Flood Explained."

11. Wallace-Wells, *Uninhabitable Earth*, 11–13. As Wallace-Wells notes, the last time the earth was warmer by four degrees Celsius, palm trees grew in the Arctic. Even if every country on the planet implements all of the commitments made in the Paris accords, we are still likely to get about 3.2 degrees of warming.

12. Between 1980 and 2015, Texas had more billon-dollar weather and climate disasters than anywhere in America—tornados, ice storms, haboobs, wind storms, heat waves, hurricanes, blizzards, droughts, and floods. It's not difficult to conclude that climate change is loading the dice against Texas. Yet Texas still has not done much to mitigate flood risk, droughts, and hurricanes.

13. Davenport, "Major Climate Report Describes."

14. Wright, "America's Future Is Texas."

15. To be situated means rhetoric is an everyday practice that always emerges with a specific context that is historically, socially, culturally, politically, discursively, contingent. As it is used here, public is not Habermasian public spheres but rather an addressivity and relationality to what is shared or common. Material-semiotics means rhetoric is never just discourse but rather the ways in which materialities, discourse, and practice are always co-productive. See chapter 1 for a more detailed discussion of rhetoric.

16. Druschke, "Trophic Future"; McGreavy et al., *Tracing Rhetoric and Material Life*;

Olson, *Constitutive Visions*; Pezzullo, *Toxic Tourism*; Rice, *Distant Publics*; Rai, *Democracy's Lot*.

17. De/colonial is an attempt to reckon with coloniality and acknowledge decolonial possibilities. I use it here to reckon with complicity in colonial practices that connects and differentially effects everyone, while also acknowledging alternatives. See Wanzer-Serrano, *New York Young Lords*; Anzaldúa, *Borderlands/La Frontera*; Anzaldúa and Keating, *This Bridge Called Home*; Cortez, "Of Exterior and Exception," 124–50; García and Baca, *Rhetorics Elsewhere and Otherwise*; Ríos, "Cultivating Land-Based Literacies and Rhetorics"; Jackson, "Decolonizing Place and Race," 292–302.

18. Haraway, "Awash in Urine," 301–16. My notion of climatic response-ability comes from Haraway's notion of response-ability as a praxis of care and response "in ongoing multispecies worlding on a wounded terra," 302.

19. Cadena and Blaser, *A World of Many Worlds*; Harding, "One Planet, Many Sciences," 39–62; Mignolo and Porter, *On Decoloniality*.

20. Cadena and Blaser, *A World of Many Worlds*, 4; Reiter, *Constructing the Pluriverse*.

21. Quijano, "Coloniality and Modernity/Rationality," 168–78; Mignolo, *Local Histories/Global Designs*; Escobar, *Designs for the Pluriverse*, 209.

22. Haraway, *Staying with the Trouble*; Stengers, "It's Matters of Concern All the Way Down."

23. Prevail comes from the Latin root *praevalēre*, to "have greater power."

24. Burke, *Rhetoric of Motives*; Charland, "Constitutive Rhetoric," 133–50; Haraway, "Promises of Monsters"; Herndl and Licona, "Shifting Agency," 133–54; Graham, *Politics of Pain Medicine*.

25. Steinberg, *Acts of God*.

26. Steinberg, *Acts of God*, xix–xx.

27. Hesford, Licona, and Teston, *Precarious Rhetorics*, 3.

28. Ben Taub, "Inequality and Hurricane Harvey."

29. Wallace-Wells, *Uninhabitable Earth*, xx.

30. Anzaldúa, *Borderlands/La Frontera*; Mignolo, *Darker Side of Western Modernity*.

31. DeChaine, *Border Rhetorics*; Cisneros, *Border Crossed Us*.

32. Mignolo, *Local Histories/Global Designs*.

33. Cintrón, *Angel's Town*; Certeau, *Practice of Everyday Life*.

34. Justin Gillis, "Prophet of Doom Was Right about the Climate," *New York Times*, June 23, 2018. www.nytimes.com.

35. Hulme, *Weathered*, 2.

36. Hulme, *Weathered*.

37. Cushman et al., "Decolonizing Projects," 1–22; Wanzer-Serrano, *New York Young Lords*. It's worth pointing out that are there are tensions and incompatibilities among cultural/Indigenous rhetorics and de/coloniality. De/coloniality is a broad political analytic, and it is particularly insightful for questioning the foundations of modernity. Cultural rhetorics were founded on cultural studies (a European phenomenon) and the cultural practices of specific cultural communities that are distinct/unique in rhetorical practice. These subareas often overlap, and they certainly do in this project. However, it's important to not conflate the two. I'd like to thank an anonymous reviewer of my

RSQ article based on chapter 3 for pointing this out to me. Unfortunately, though I had heard of Cushman's work on decoloniality, I had not read her "Decolonizing Projects" article before writing this book. It was only during the peer review process that I was made aware of this particular article, which does pair well with Wanzer-Serrano's orientation to de/coloniality. I attribute this oversight to the differences among composition/literacy studies and rhetorical studies, which can overlap but often conference and publish in different venues. I choose to see this as a strength of rhetoric and composition studies. For arguments about how and why cultural rhetorics are a form of decoloniality, see Powell et al., "Our Story Begins Here."

38. Tuck and Yang, "Decolonization Is Not a Metaphor." I address some of these concerns in the preface and return to them more fully in the coda, but the practice behind this ethic is woven throughout the book.

39. Harding, "One Planet, Many Sciences," 39–62.

40. Wanzer-Serrano, *New York Young Lords*, 11. More specifically, I draw here on Wanzer-Serrano's concept of de/coloniality to emphasize a theoretical perspective attentive to coloniality that may or may not also include reference to decolonial options. De/coloniality as an analytic holds much promise for rhetorical studies. I address this topic in detail in the first chapter.

41. Cadena and Blaser, *World of Many Worlds*, 4.

42. Harding, *Sciences from Below*; Harding, "State of the Field," 48–63.

43. Escobar, *Designs for the Pluriverse*; Reiter, *Constructing the Pluriverse*.

44. Intergovernmental Panel on Climate Change (IPCC), "Summary for Policy Makers."

45. Bretcher, "Making the Green New Deal Work for Workers."

46. Dove and Carpenter, *Environmental Anthropology*.

47. Roberts, "Hurricane Katrina Showed."

48. Klinenberg, "Want to Survive Climate Change?"

49. Klein, *This Changes Everything*.

50. Jamie Hopkins and Kiah Collier, "Surge of Oil and Gas Flowing to Texas Coastline Triggers Building Boom, Tensions," *Texas Tribune*, November 29, 2018, www.texastribune.org.

51. Hawhee and Olson, "Pan-Historiography," 90–105; Rai and Druschke, *Field Rhetoric*.

52. Olson, *Constitutive Visions*, xx.

53. Ackerman, "Space for Rhetoric in Everyday Life"; McGreavy et al., *Tracing Rhetoric and Material Life*; McKinnon et al., *text + FIELD*; Rai and Druschke, *Field Rhetoric*; Rai, *Democracy's Lot*.

54. McKinnon et al., *text + FIELD*, 5–6.

55. Rosales, *Illusion of Inclusion*.

56. Santos, "San Antonio Is a City of Metamorphosis."

57. Miller, "Genre as Social Action," 151–76; Spinuzzi, *Tracing Genres through Organizations*.

58. McKinnon et al., *text + FIELD*; Rai and Druschke, *Field Rhetoric*.

Chapter One

1. Reiter, *Constructing the Pluriverse*, ix.

2. Quijano, "Coloniality and Modernity/Rationality," 168–78.

3. Cadena and Blaser, *World of Many Worlds*, 4.

4. Burke, *Grammar of Motives*.

5. Onís, "Looking Both Ways," 308–27; Haas and Frost, "Toward an Apparent Decolonial Feminist Rhetoric of Risk," 168–86; Herndl and Licona, "Shifting Agency," 133–54; McGreavy et al., *Tracing Rhetoric and Material Life*; Pezzullo, *Toxic Tourism*.

6. Anzaldúa, *Borderlands/La Frontera*; Cisneros, *Border Crossed Us*; DeChaine, *Border Rhetorics*; Mignolo, *Darker Side of Western Modernity*.

7. Pezzullo, *Toxic Tourism*; Sackey, "An Environmental Justice Paradigm for Technical Communication," 138–62.

8. Druschke, "Trophic Future"; Edbauer, "Unframing Models of Public Distribution," 5–24; Eubanks, *Troubled Rhetoric and Communication of Climate Change*; Herndl and Brown, *Green Culture*; Killingsworth and Palmer, *Ecospeak*; Rivers, "Deep Ambivalence and Wild Objects," 420–40; Rickert, *Ambient Rhetoric*; Ross, *Topic-Driven Environmental Rhetoric*; Simmons, *Participation and Power*; Walsh, *Scientists as Prophets*; Walker, "Mapping the Contours of Translation," 104–20.

9. McGreavy et al., *Tracing Rhetoric and Material Life*; Peterson, *Sharing the Earth*; Pezzullo, *Toxic Tourism*; Sandler and Pezzullo, *Environmental Justice and Environmentalism*; Sowards, "Environmental Justice in International Contexts," 37–41; Stormer and McGreavy, "Thinking Ecologically about Rhetoric's Ontology," 1–25.

10. McGreavy et al., *Tracing Rhetoric and Material Life*, 4.

11. RSTM has notable strengths at the intersections of gender studies (Condit, *Meanings of the Gene*; Jack, *Science on the Home Front*; Keränen, *Scientific Characters*); disability studies (Jack, *Autism and Gender*; Johnson, *American Lobotomy*); and new materialism (Graham, *Politics of Pain Medicine*), to a name a few.

12. Happe, *Material Gene*.

13. Pezzullo, *Toxic Tourism*; Onís, "Looking Both Ways"; McGreavy et al., *Tracing Rhetoric and Material Life*. One notable exception is the early work of Carl Herndl, who more than any other environmental rhetorician embraced critical cultural theories before turning his attention to new materialisms.

14. Place-based, community engaged, and field methodologies from ecology and RSTM have done the most productive work in this regard. I take up critical emplaced rhetorical work explicitly in the next section.

15. Scott, "Extending Rhetorical-Cultural Analysis," 351–53; Scott, *Risky Rhetoric*.

16. Blake Scott is more materialist than new materialist, and it's important to understand those distinctions, which mostly have to do with the nature of language and experience. For a helpful review of the new materialist turn in rhetoric of science, technology, and medicine, see Graham, *Politics of Pain*, and Booher and Jung, *Feminist Rhetorical Science Studies*.

17. Booher and Jung, *Feminist Rhetorical Science Studies*; Graham, *Politics of Pain*.

18. Haraway, *Staying with the Trouble*; Stengers, *Another Science Is Possible*.

19. Durability and portability are Latourian terms that describe how scientific knowledges are recalcitrant to change and therefore portable among other disciplinary communities. In addressing coloniality head-on, rhetoric of science might be differently durable and portable to subdisciplines of rhetoric and critical cultural studies that tend to eschew rhetorical science studies writ large. Latour, *Science in Action*.

20. Scott, "Extending Rhetorical-Cultural Analysis," 356–57.

21. Haraway, *Staying with the Trouble*; Scott, "Extending Rhetorical-Cultural Analysis."

22. Latour, *Science in Action*.

23. Druschke and McGreavy, "Why Rhetoric Matters for Ecology," 46–52.

24. Goodwin, *Between Scientists and Citizens*; Druschke et al., "Better Science through Rhetoric," 175–90.

25. Herndl et al., "What's a Farm?," 61–94; Graham, *Politics of Pain*.

26. Goodwin et al., "Rhetorical Resources for Teaching Responsible Communication of Science"; Mehlenbacher, *Science Communication Online*; Walker, "Rhetorical Principles on Uncertainty," 1–13.

27. Graham, "Promise and Peril of Scientific Science Studies," 529–53.

28. Sandoval, *Methodology of the Oppressed*; Graham, "Promise and Peril"; Frost, *Biocultural Creatures*.

29. Frost, *Biocultural Creatures*.

30. Frost, *Biocultural Creatures*, 557–58.

31. Garcia and Baca, *Rhetorics Elsewhere and Otherwise*.

32. The nonmodern as a root for both science studies and de/coloniality is a point I address more specifically through divergence and diplomacy in chapter 3.

33. Mignolo quoted in Wanzer-Serrano, *New York Young Lords*, 75.

34. Wanzer-Serrano, *New York Young Lords*, 15.

35. Wanzer-Serrano, *New York Young Lords*, 15.

36. Booher and Jung, *Feminist Rhetorical Science Studies*.

37. Sandoval, "New Sciences."

38. Sandoval, *Methodology of the Oppressed*.

39. Haraway, *Staying with the Trouble*.

40. Wanzer-Serrano, *New York Young Lords*, 183.

41. Sandoval, *Methodology of the Oppressed*, 176.

42. Druschke, "Trophic Future"; McGreavy et al., *Tracing Rhetoric and Material Life*; Stormer and McGreavy, "Thinking Ecologically"; Rivers, "Deep Ambivalence"; Rickert, *Ambient Rhetoric*.

43. Herndl and Cutlip, "How Can We Act?"; Scott, *Politics of Pain*.

44. Druschke, "Trophic Future."

45. Cadena and Blaser, *World of Many Worlds*, 9.

46. Boyle, *Rhetoric as a Posthuman Practice*.

47. Stengers, "Including Nonhumans," 3–31.

48. Stengers, "Including Nonhumans"; Haraway, *Staying with the Trouble*.

49. Haraway, "Situated Knowledges," 575–99.

50. Druschke, "Trophic Future"; Cadena, *Earth Beings*.

51. Stengers, "Including Nonhumans," 24.

52. Stengers, "Including Nonhumans," 25.

53. Stengers, "Including Nonhumans," 25.

54. Stengers, "Including Nonhumans," 28.

55. Stengers, "Including Nonhumans," 20.

56. Stengers, "Cosmopolitical Proposal," 995.

57. Stengers, "Introductory Notes on an Ecology of Practices," 186.

58. Stengers, "Cosmopolitical Proposal," 1002.

59. Stengers, "Cosmopolitical Proposal," 1002.

60. Stengers, "Cosmopolitical Proposal," 1003.

61. Cadena and Blaser, *World of Many Worlds*, 14.

62. Stengers, "Challenge of Ontological Politics," 91.

63. Stengers, "Challenge of Ontological Politics," 91.

64. Druschke, "Trophic Future"; Keeling and Prairie, "Trophic and Tropic Dynamics," 39–58; Druschke and Rai, "Making Worlds with Cyborg Fish," 197–222.

65. Stengers, "It's Matters of Concern."

66. Cadena and Blaser, *World of Many Worlds*, 4.

67. Latour, *We Have Never Been Modern*. Latour has made it clear that it was never his intention to address these issues, but ending such silence will help achieve a more entangled version of the nonmodern.

68. Mignolo, *Darker Side of Western Modernity*, 274.

69. Mignolo, *Darker Side of Western Modernity*, 274.

70. Mignolo, *Darker Side of Western Modernity*, 274, 293–94.

71. Asen, "Discourse Theory of Citizenship," 189–211; Rai, *Democracy's Lot*; Rice, *Distant Publics*.

72. Ackerman, "Space for Rhetoric in Everyday Life"; Rai and Druschke, *Field Rhetorics*, 3.

73. Rai, *Democracy's Lot*, 6.

74. Pezzullo, *Toxic Tours*; Rai and Druschke, *Field Rhetoric*.

75. Brouwer and Asen, *Public Modalities*. Everyday and vernacular rhetorics correct for Habermasian notions of the public sphere and public deliberation in numerous ways, but particularly by acknowledging the profound situatedness of rhetoric that is attuned to the local, marginalized, countercultural, ecological, and material.

76. Greene quoted in Rai, *Democracy's Lot*, 14.

77. Undoubtedly, some scholars will find some of these terms inherently problematic given their historical associations with Greco-Roman rhetoric. Yet, we should note that these concepts have their earliest known developments in Sophistic rhetoric, and they are by design adaptable across contexts while also providing an analytic for language practices. The earliest notion of *topoi* as commonplaces of discourse does not come from Aristotle but from the Sophists and a cosmopolitan culture widely influenced by migrations across North Africa, Asia, and Europe.

78. For useful reviews, see Boyle, *Rhetoric as a Posthuman Practice*; Boyle and Walsh, *Topologies as Techniques*; Rai, *Democracy's Lot*; Walker and Walsh, "No One Knows What the Ultimate Consequences May Be," 3–34; and Walsh, "Common Topoi of STEM Discourse," 120–56.

79. Rai, *Democracy's Lot*, 38–42; Muckelbauer, *Future of Invention*.

80. Boyle, *Rhetoric as a Posthuman Practice*, 135, emphasis in the original.

81. Walsh and Boyle, *Topologies as Techniques*, 2.

82. With roots in the Greek word for season, *kairos* in the context of climate captures the fleeting and contingent opportunities for change made material by the relationship between climate and weather. In this sense, *topos* and *kairos* discursively figure the relationship between climate and weather, between stability and change, and between the return and the circulation of important political rhetorics.

83. Walsh and Boyle, *Topologies as Techniques*, 2.

84. Walsh and Boyle, *Topologies as Techniques*, 5; Walsh, "Resistance and Common Ground," 458–87.

85. Fahnestock, *Rhetorical Figures in Science*.

86. Burke, *Philosophy of Literary Form*; Burke, *Grammar of Motives*; Harris, "Dementia, Rhetorical Schemes, and Cognitive Resilience."

87. Muckelbauer, "Implicit Paradigms of Rhetoric," 30–41.

88. Druschke, "Trophic Future"; Keeling and Prairie, "Trophic and Tropic Dynamics"; see also Abeles, Jack, and Singer, "Resilient Turns."

89. Rickert, *Ambient Rhetoric*; Keeling and Prairie, "Trophic and Tropic Dynamics."

90. Anzaldúa, *Borderlands/La Frontera*, vii–3.

91. Similarly, what geohumanist Nishat Awan calls border topologies highlight an understanding of borders as spatial and ecological entities that de-link borders from territory and from an understanding of them as a technology of separation. See Awan, "Introduction to Border Topologies," 279–83.

92. Rice, *Distant Publics*; Rickert, *Ambient Rhetoric*.

93. Foucault, *Order of Things*.

94. Reeves, "Suspended Identification," 306–27; Cortez, "Of Exterior and Exception," 124–50.

95. Cadena, *Earth Beings*.

96. Olson, *Constitutive Visions*, 12–14.

97. Burke, *Grammar of Motives*; Stormer and McGreavy, "Thinking Ecologically," 325.

98. Hawk, *Resounding the Rhetorical*, 177.

99. In Tlostanova and Mignolo, "On Pluritopic Hermeneutics, Trans-modern Thinking, and Decolonial Philosophy," 16.

100. Tlostanova and Mignolo, "On Pluritopic Hermeneutics, Trans-modern Thinking, and Decolonial Philosophy," 18.

101. Escobar, *Designs for the Pluriverse*, 209, emphasis mine.

102. Rickert, *Ambient Rhetoric*, 49.

103. Mignolo, *Local Histories/Global Designs*.

104. Mignolo, *Darker Side of Western Modernity*, 208.

105. Sandoval, *Methodology of the Oppressed*, 176.

106. Anzaldúa, *Borderlands/La Frontera*; Moraga and Anzaldúa, *This Bridge Called My Back*. Anzaldúa and Keating, *This Bridge Called Home*; Anzaldúa, *Luz en lo Oscuro/ Light in the Dark*.

107. Anzaldúa, *Borderlands/La Frontera*; Chávez, *Queer Migration Politics*; Licona, *Zines in Third Space*; Sandoval, *Methodology of the Oppressed*.

108. Baca, *Mestiz@ Scripts*; Medina, *Reclaiming Poch@ Pop*; Mejía, "Tejano Arts of the U.S.-Mexico Contact Zone," 123–35; Ramírez, *Occupying Our Space*.

109. DeChaine, *Border Rhetorics*.

110. Baca and Villanueva, *Rhetorics of the Americas*; Cortez, "Of Exterior and Exception"; De Los Santos, "The Future of Our History," 199–211; Olson, *Constitutive Visions*; Olson and De Los Santos, "Expanding the Idea of América," 193–98.

111. Cisneros, *Border Crossed Us*, 143.

112. Cisneros, *Border Crossed Us*, 161–62.

113. Mignolo, *Local Histories*, 11.

114. Mignolo, *Local Histories*, 18.

115. Anzaldúa, *Borderlands/La Frontera*; Cisneros, *Border Crossed Us*; Mignolo, *Darker Side of Western Modernity*.

116. Saldívar, *Border Matters*, 32.

117. Cisneros, *Border Crossed Us*, 32.

118. Garcia and Baca, *Rhetorics Elsewhere and Otherwise*, 2.

119. As I show throughout, quite independently nonmodern scholars, including rhetorical scholars, have also offered alternatives to these dualities via engaged, praxiographic, and ecological orientations. See Graham, *Politics of Pain*; McGreavy et al., *Tracing Rhetoric and Material Life*.

120. Wanzer-Serrano, *New York Young Lords*, 12.

121. Mignolo, *Darker Side of Western Modernity*, 293.

122. Mignolo, *Darker Side of Western Modernity*, 293; Harding, "One Planet, Many Sciences"; Harding, "State of the Field."

123. Cadena and Blaser, *World of Many Worlds*.

124. Harding, "Latin American Decolonial," 1063–87.

125. Escobar, *Designs for the Pluriverse*.

126. Escobar, *Designs for the Pluriverse*, ix.

127. Escobar, *Designs for the Pluriverse*, ix.

128. Escobar, *Designs for the Pluriverse*, ix.

129. Wanzer-Serrano, *New York Young Lords*; Escobar, *Designs for the Pluriverse*, 149.

130. Harding, "Latin American Decolonial," 1078.

131. Bullard, *Dumping in Dixie*.

132. For an example, many scholars point to the First National People of Color Environmental Leadership Summit, which produced the Principles of Environmental Justice in 1991.

133. Peña, *Mexican Americans and the Environment*, 153.

134. Bullard, *Dumping in Dixie*, 13; See also Pezzullo, *Toxic Tours*; Pezzullo and Sandler, *Environmental Justice and Environmentalism*.

135. Black Congressional Caucus, *African Americans and Climate Change*; Schlosberg and Collins, "From Environmental to Climate Justice," 362.

136. Cintrón, "Abandon All Hope Ye," 189–212; Escobar, *Designs for the Pluriverse*.

137. Harding, "One Planet, Many Sciences," 48–52.

138. Tlostanova and Mignolo, "On Pluritopic Hermeneutics," 18.

139. Mignolo, *Local Histories*.

140. Wanzer-Serrano, *New York Young Lords*, 180–81.

Chapter Two

1. "Hot Summers Long a Way of Life," *San Antonio Express News*, August 28, 2018. This article notes: "San Antonio has seen a steep rise over the past 130 years in the number of extremely hot days per year, according to NOAA data. From 1890 to 1899, San Antonio recorded 38 days of 100 degrees or more. By the 1950s, that had risen to 124 days. After a dip to only seven days of 100 degrees or higher in the 1970s—a relatively cool decade worldwide—the mercury began routinely rising again. There were 101 days in the 1980s, 143 in the 1990s and 147 in the 2000s. In the 2010s, with two summers left to go, San Antonio already has sweated through 178 days of 100 degrees or hotter—with four of those coming earlier this year [2018]."

2. Local climatologists predict a 7 percent intensification of precipitation due to an increase in evaporation. Hayhoe, "Climate Trends in San Antonio," 5.

3. Flynn, Sotirin, and Brady, *Feminist Rhetorical Resilience*; McGreavy, "Resilience as Discourse," 104–21. Walker and Cagle, "Resilience Rhetorics in Science, Technology, and Medicine."

4. Cadena and Blaser, *World of Many Worlds*.

5. In Mignolo, "Foreword: On Pluriversality and Multipolarity."

6. Flynn, Sotirin, and Brady, *Feminist Rhetorical Resilience*, 8.

7. Flynn, Sotirin, and Brady, *Feminist Rhetorical Resilience*, 12.

8. Bean, Keränen, and Durfy, "This Is London," 427–64; Charland, "Constitutive Rhetoric," 133–50; Paliewicz, "Bent but Not Broken," 1–14. Though often discussed as a rhetorical version of interpellation, constitutive rhetorics can be viewed as an early ecological rhetoric wherein "rhetoric works not by persuading fixed subjects, but in the potent flux of ideological forces that prompt the constitution of subjects in the first place." See McGreavy et el., *Tracing Rhetoric and Material Life*, 9.

9. Bean, Keränen, and Durfy, "This Is London," 453.

10. Stormer and McGreavy, "Thinking Ecologically," 1–25.

11. Stormer and McGreavy, "Thinking Ecologically," 19.

12. McGreavy, "Resilience as Discourse," 115.

13. Hesford, Licona, and Teston, *Precarious Rhetorics*.

14. Hesford, Licona, and Teston, *Precarious Rhetorics*, 6.

15. Liévanos, "Minority Perspective Is Limited," 1; Norgaard, *Living in Denial*; Pulido, "Rethinking Environmental Racism," 12–40.

16. Anguiano et al., "Connecting Community Voices," 124–43.

17. Norgaard, *Living in Denial*, 222.

18. Portions of this work have been featured in the Texas Humanities–funded local web series, *Hidden Histories*: "From Climate Strike to Climate Thunder: Unearthing Environmental Resistance in South Texas," *Urban-15: Hidden Histories*, September 29, 2019. https://urban15.org.

19. Martinez, *Injustice Never Leaves You.*

20. Hernandez-Ehrisman, *Inventing the Fiesta City.*

21. Hernandez-Ehrisman, *Inventing the Fiesta City,* 59.

22. Montejano, *Anglos and Mexicans.*

23. Sandoval, *Our Legacy,* 4–7.

24. Montejano, *Anglos and Mexicans.*

25. Lawhn, "María Luisa Garza," 83–96.

26. Donecker, "San Antonio Light," *Handbook of Texas Online.*

27. Donecker, "San Antonio Express News," *Handbook of Texas Online.*

28. Buitron, *Quest for Tejano Identity,* 19.

29. Madero was later overthrown by a coup known as the Ten Tragic Days, a coup that the American ambassador Henry L. Wilson helped organize.

30. Buitron, *Quest for Tejano Identity,* 19.

31. Buitron, *Quest for Tejano Identity.*

32. Buitron, *Quest for Tejano Identity,* 20.

33. Buitron, *Quest for Tejano Identity,* 22.

34. Yet for all of its emphasis on Mexican culture and politics, the success of *La Prensa* also "marked an important transitional period from the first generation of Mexican immigrants to their children, whose intellectual attention was now focused on seeking identity as Americans." See Buitron, *Quest for Tejano Identity,* 22.

35. Lawhn, "María Luisa Garza," 84.

36. Fisher, *American Venice,* 55.

37. Fisher, *American Venice,* 55.

38. Miller, "Streetscape Environmentalism," 158–77.

39. Miller, "Streetscape Environmentalism," 169.

40. "Officers Describe Terror of Districts Struck by Calamity," *San Antonio Light,* 41, no. 234, September 10, 1921. 2.

41. Hesford, Licona, and Teston, *Precarious Rhetorics,* 2018.

42. Montejano, *Anglos and Mexicans;* Zimring, *Clean and White.*

43. "Mexicans Lose All on the Alazan," *San Antonio Express,* 56, no. 250, September 12, 1921, 12.

44. "Mexicans Lose All," 12.

45. "I Am the Spirit of San Antonio," *San Antonio Evening News,* 4, no. 7, September 10, 1921, 1.

46. Stormer and McGreavy, "Thinking Ecologically," 16.

47. "Saturday Flood a Sequel to One of Hundred Years Ago," *San Antonio Light,* 41, no. 236, September 12, 1921, 9.

48. All translations have been completed by certified American Translators Association Member #455136.

49. Though I discuss *La Tragedia* as though Quiroga was its author, it is clear that he was only the editor of this book, and it is likely that many journalists contributed to the project. However, the names of these journalists are lost, partially because of the journalistic convention at the time to not publish author names.

50. Orozco, *No Mexicans, Women, or Dogs Allowed,* 58. These cultural clashes were

reborn a generation later in the Chicana/o movement's characterization of the Order Sons of America (OSA) and the League of United Latin American Citizens (LULAC) as assimilationists.

51. Lyon, *Deliberative Acts*, 140–41.

52. Libreria de Quiroga, *La Tragedia de la Inundacion de San Antonio*, 2.

53. Quiroga, *Tragedia*, 8–10.

54. Quiroga, *Tragedia*, 60.

55. Quiroga, *Tragedia*, 64.

56. Just for one example, sacrifice can be traced back to Mexican Indigenous and Christian religious traditions that are both based on notions of a blood sacrifice to achieve redemption and salvation.

57. Quiroga, *Tragedia*, 64.

58. Allen, *Talking to Strangers*.

59. Quiroga, *Tragedia*, 46.

60. Quiroga, *Tragedia*, 49.

61. Lyon, *Deliberative Acts*, 23.

62. Quiroga, *Tragedia*, 49.

63. Lyon, *Deliberative Acts*, 23.

64. Flood Relief Committee, *Memorandum to the San Antonio Board of Engineers*.

65. Miller, "Streetscape Environmentalism," 13.

66. Miller, "Streetscape Environmentalism," 12.

67. Orozco, *No Mexicans, Women, or Dogs Allowed*, 58.

68. The OSA largely focused on civil rights issues like ending slavery conditions, promoting workers' safety and fair wages, abolishing child labor, desegregating schools, disseminating information broadly, and maintaining Sunday as a day of rest. However, their emphasis on quality of life conditions and community health protections clearly links their work to environmental concerns.

69. Orozco, *No Mexicans, Women, or Dogs Allowed*, 232–33.

70. The inclusion of Mexican nationals in these early civil rights organizations was controversial, yet their organization also had an agenda for "the advancement, progress, and prosperity of the people of our extraction in general, *regardless of citizenship*" (Orozco, *No Mexicans, Women, or Dogs Allowed*, 231, italics mine). Thus, even in the earliest turns toward US citizenship as a tool for political power, Mexicano cultural solidarity and restitution remained closely tied together, as they did in Quiroga's writing.

Chapter Three

1. "Climate Mayors," accessed January 5, 2017, http://climatemayors.org/.

2. Steinberg, *Acts of God*.

3. Cadena and Blaser, *World of Many Worlds*, 4; Mignolo and Porter, *On Decoloniality*.

4. Klinenberg, "Want to Survive Climate Change?" Klinenberg's research shows that it is the presence of a social infrastructure in a neighborhood that turns out to be a major predictor of who lives and dies during extreme weather events, in communities both rich and poor, where "neighbors are the first true responders."

5. Climate breakdown is a term George Monbiot coined in the wake of the IPCC's

release of AR4 to describe how the climate variability, extreme weather events, and general climate conditions that fostered human/nonhuman evolution on this planet is now undergoing a collapse. George Monbiot, "Climate Change? Try Catastrophic Climate Breakdown," last modified September 27, 2013, *The Guardian*, www.theguardian.com.

6. Cadena and Blaser, *World of Many Worlds*.

7. Stengers, "Challenge of Ontological Politics," 85.

8. Chávez, *Queer Migration Politics*, 7.

9. Licona and Chávez, "Relational Literacies," 96–107.

10. Cisneros, *Border Crossed Us*; Cadena and Blaser, *World of Many Worlds*; Garcia and Baca, *Rhetorics Elsewhere and Otherwise*; Mignolo, *Local Histories/Global Designs*.

11. Anzaldúa, *Borderlands/La Frontera*; Cisnernos, *Border Crossed Us*; Mignolo, *Local Histories/Global Designs*; Reiter, *Constructing the Pluriverse*.

12. Stormer, "Articulation," 257–84.

13. Stengers, "Including Nonhumans in Political Theory," 25. In this sense, Stengers's version of *oikos* is related to Grecian farms as a basic unit of the agricultural economy, which was also dependent on slavery. Though that relation is not necessary. By insisting that ethos and *oikos* are inseparable and share relationships, they can function as speculative heuristics for alternative developments and transitions.

14. Stengers, "Cosmopolitical Proposal," 995. Cosmopolitics is not Kantian. It is also not cosmopolitanism. The relationship between cosmopolitics and cosmopolitanism is outside the scope of this analysis, but for useful examples see Watson, "Derrida, Stengers, Latour, and Subalternist Cosmopolitics."

15. Stengers, "Introductory Notes on an Ecology of Practices," 186.

16. Chávez, *Queer Migration Politics*.

17. Stengers, "Cosmopolitical Proposal," 1002.

18. Chávez, *Queer Migration Politics*.

19. Stengers, "Including Nonhumans," 28.

20. Stengers, "Including Nonhumans," 29.

21. Stengers, "Including Nonhumans," 27.

22. Stormer, "Articulation"; Watson, "Derrida, Stengers, Latour, and Subalternist Cosmopolitics," 75–98. As Matthew Watson has identified, Stengers's lifelong project is to show how what we characterize as "the modern sciences" were never modern at all. To do this, Stengers highlights the constraints of scientific practice and how its truth-claims emerge out of a specific configuration of material systems that have no universal bearing on other worlds without those configurations. Thus, Stengers de-links the practice of science from its colonial/modern claims to universality. In analyzing scientific practice and its institutions with an overtly political language, and in emphasizing divergence not contradiction, Stengers "opens up the possibility that ostensibly contradictory scientific programs or metaphysical systems can coexist peacefully, affirming their independent logics of composition and knowledge production," Watson, 87.

23. Latour, *Inquiry into Modes of Existence*.

24. Herndl and Graham, "Getting Over Incommensurability," 40–58.

25. In Walsh et al., "Forum: Bruno Latour and Rhetoric," 403–62.

26. María Antonietta Berriozábal, personal correspondence, May 2, 2019.

27. Sandoval, *Methodology of the Oppressed.*

28. Cadena, *Earth Beings.* I'd like to thank Caroline Gottschalk Druschke for bringing my attention to the work of Cadena. See Druschke, "Trophic Future for Rhetorical Ecologies."

29. Cadena, *Earth Beings*, 280.

30. Cadena, *Earth Beings*, 282.

31. Cadena, *Earth Beings*, 281.

32. Mignolo, *Darker Side of Western Modernity*, 207.

33. Though it's beyond the scope of this analysis, de-linking and divergence clearly have some overlaps. But a beginning distinction between them would acknowledge their different histories and potentialities. In the context of this analysis, de-linking is more of a theoretical concept about onto-epistemologies that are otherwise, while divergence is radically immanent.

34. Wanzer-Serrano, *New York Young Lords*, 16. Wanzer-Serrano and others have noted a significant alignment among decolonial analysis and social movements. From rhetorical scholarship, this relationship has been most fully articulated by Wanzer-Serrano, who used decolonial theory to rhetorically analyze the Young Lords in New York City during the 1960s. Decoloniality and environmental/climate justice movements are not completely compatible. Their genealogies are quite different. Their alignments are most thoroughly developed in studies of the Global South (Álvarez and Coolsaet, "Decolonizing Environmental Justice Studies"; Rodríguez and Inturias, "Conflict Transformation"), but they also extend to environmental justice movements of the southern United States in the 1970s, which arose out of the lived experiences of people of color facing systematic oppressions like the siting of toxic waste sites in their backyards (see Bullard, *Dumping in Dixie*; Pulido, "Rethinking Environmental Racism").

35. Cadena and Blaser, *World of Many Worlds*, 9.

36. Stengers, "Challenge of Ontological Politics," 91.

37. Anzaldúa and Keating, *This Bridge We Call Home*, 3.

38. Licona and Chávez, "Relational Literacies and Their Coalitional Possibilities," 98; Herndl and Licona, "Shifting Agency," 133–54.

39. Lugones, *Pilgrimages/Peregrinajes*, ix.

40. Chávez, *Queer Migration Politics*, 5–7.

41. Chávez, *Queer Migration Politics*, 9.

42. Mignolo and Porter, *On Decoloniality: Concepts, Analytics, Praxis*, 155.

43. Stengers, "Challenge of Ontological Politics," 91.

44. Haraway, "Promises of Monsters."

45. Rogers, *Cold Anger.*

46. Brischetto, Cottrell, and Stevens, "Conflict and Change," 75–94.

47. Berriozábal, *María, Daughter of Immigrants*; María Antonietta Berriozábal, personal correspondence, May 2, 2019.

48. Montejano, *Quixote's Soldiers*, 148.

49. Montejano, *Quixote's Soldiers*, 209. The University of Texas, San Antonio, was created in 1969 because of a successful lawsuit by the Mexican-American Legal Defense and Education Fund (MALDEF), who argued for the campus based on the need for

Chicana/os access to higher education. The state reluctantly created the branch campus, but sited it thirty minutes from the city center in the middle of a majority Anglo and middle-class suburb.

50. Montejano, *Quixote's Soldiers*, 210.

51. Rogers, *Cold Anger*, 111.

52. Kelly, "Détournement, Decolonization," 168–90. COPS is founded on three closely intertwined mediating institutions: first, the family—whose primary concerns are good jobs, a clean and healthy environment to live and play in, and a quality education for their children; second, the neighborhood—whose concern was quality of life, affordable housing, and an identifiable culture; third, the church—whose obligation is to promote social justice for the poor, and in the tradition of liberation theology, to protect family and neighborhoods from poverty (see Sekul, "Communities Organized for Public Service," 177). COPS has an elaborate structure designed specifically to keep a check on powerful individuals.

53. For these reasons, COPS can be appropriately situated at the intersections of multiple areas of rhetorical study—classical rhetoric, rhetorical pragmatism, Chicana/o rhetorics, social movements, religious rhetorics, and environmental rhetorics.

54. Rivers, "Deep Ambivalence and Wild Objects," 420–40; Alinsky, *Rules for Radicals*.

55. Danisch, *Pragmatism, Democracy, and the Necessity of Rhetoric*.

56. Cisneros, *Border Crossed Us*.

57. Mignolo, *Local Histories/Global Designs*.

58. Stengers, "Including Nonhumans."

59. Ray Kaiser, Press Release, Special Collections, University of Texas, San Antonio.

60. Jarboe, "Building a Movement," 38–46.

61. Jarboe, "Building a Movement," 44; Miller, "Streetscape Environmentalism," 158–77.

62. Miller, "Streetscape Environmentalism," 174.

63. I'd like to thank Catherine Chaput for bringing my attention to the concept of fleshy subjectivity.

64. Lugones, *Pilgrimages/Peregrinajes*.

65. Rogers, *Cold Anger*, 123.

66. Rogers, *Cold Anger*, 123.

67. Rogers, *Cold Anger*, 124.

68. Stengers, "Including Nonhumans," 27.

69. Cisneros, *Border Crossed Us*; Cadena, *Earth Beings*.

70. Andy Sarabia, Press Release, Special Collections, the University of Texas, San Antonio.

71. The story of the APA began when the San Antonio branch of the League of Women Voters grew increasingly concerned about how suburban development would negatively affect water quality. They asked one of their members, a woman by the name of Fay Sinkin, to start an organization that could help protect San Antonio's most precious resource—it's fantastically clean source of water, the Edwards Aquifer. It's worth noting that as organizations, the League of Women Voters and the Communi-

ties Organized for Public Services had similar profiles—both were activist, grassroots, and nonpartisan organizations that focused on molding political leaders by informing citizens and obtaining results on specific issues of public concern.

72. Cadena and Blaser, *World of Many Worlds*, 4; Stengers, "Challenge of Ontological Politics."

73. The local experts I have talked with all agree the APA could have done more on this point, but they also point out that COPS typically did not work with any other organizations, especially single issue-driven organizations like environmental groups.

74. Plotkin, "Democratic Change," 157–74.

75. Plotkin, "Democratic Change," 171.

76. Montejano, *Quixote's Soliders*, 242.

77. Montejano, *Quixote's Soliders*, 250.

78. Cadena and Blaser, *World of Many Worlds*.

79. As one of the most effective pieces of federal legislation ever enacted in the United States, the Voting Rights Act is best thought of as a human-centered version of the pluriversal—it attempts to mitigate the violence creation of one world (Whites only), it empowers a plurality of situated public discourse (no literacy tests; bilingual ballots), and such. Indeed, though it is human-only, a benchmark for effective federal policy may very well be if it helps create pluralism, and eventually, perhaps, a world of many worlds.

80. Plotkin, "Democratic Change," 173.

Chapter Four

1. Wallace-Wells, *Uninhabitable Earth*. The lifetime of CO_2 is difficult to determine, but most estimates place it between twenty and two hundred years if it dissolves into the ocean—between 65 percent and 80 percent. The rest can be left in the atmosphere for thousands of years. Methane persists in the atmosphere for about twelve years. Nitrous oxide is broken down in the stratosphere and persists for around fourteen years. Duncan Clark, "How Long Do Greenhouse Gasses Stay in the Air?," *The Guardian*, January 16, 2012, www.theguardian.com.

2. Walker, "Rhetorics of Uncertainty."

3. McGreavy, "Resilience as Discourse," 104–21; Schlosberg and Collins, "From Environmental to Climate Justice," 359–74.

4. Bendell, "Six Months of the Deep Adaptation Forum."

5. Pelling, *Adaptation to Climate Change*.

6. Rice, *Distant Publics*; Rai, *Democracy's Lot*.

7. Bedoya, "Spatial Justice"; Lipsitz, *How Racism Takes Place*.

8. Andrew Revkin, "A River Runs under It," *New York Times: Dot Earth Blog*, accessed July 15, 2019, https://dotearth.blogs.

9. Cortez, "No Nos Moveran," 86; Logan and Molotch, *Urban Fortunes*.

10. Cortez, "No Nos Moveran," 86.

11. Rice, *Distant Publics*, 18.

12. Gries, *Still Life with Rhetoric*; Olson, *Constitutive Visions*; Walker, "Mapping the Contours of Translation," 104–20.

13. Druschke, "Trophic Future for Rhetorical Ecologies"; Graham, "Object-Oriented Ontology's Binary Duplication," 108–24.

14. Muckelbauer, "Implicit Paradigms of Rhetoric," 30–41.

15. Escobar, *Designs for the Pluriverse*.

16. Escobar, *Designs for the Pluriverse*, 211.

17. On a technical level, the county discovered that the FEMA hundred-year flood-plain measurements assumed the creek was 30 to 53 feet in width, when in reality, the creek only varied from 20 to 28 feet. It's no surprise FEMA was wrong, but it meant that most of the existing development along the creek was directly in the floodplain, and the project was supposed to contain at least one-hundred-year flood events by increasing the width and depth of the creek channel. San Pedro Creek Preliminary Engineering Report, May 16, 2013, 26.

18. Muñoz, *Muñoz*.

19. Muñoz, *Muñoz*.

20. San Pedro Creek Preliminary Engineering Report, May 16, 2013.

21. San Pedro Creek Preliminary Engineering Report, May 16, 2013.

22. Bedoya, "Spatial Justice"; Lipsitz, *How Racism Takes Place*.

23. Bedoya, "Spatial Justice."

24. Ybarra-Frausto, "Rasquachismo."

25. Bedoya, "Spatial Justice." For example, Rasquachismo aesthetics might repurpose a tire of a semitruck by painting it pink and using it as a flowerpot in the front yard—not exactly an object one would find at Lowe's. I'd like to thank Tomás Ybarra-Frausto for the conversations we have had on this topic and others.

26. Bedoya, "Spatial Justice."

27. Escobar, *Designs for the Pluriverse*; Pelling, *Adaptation to Climate Change*.

28. Escobar, *Designs for the Pluriverse*, 211.

29. Escobar, *Designs for the Pluriverse*, 217.

30. "Authenticity," *Oxford English Dictionary*.

31. Sandoval, *Methodology of the Oppressed*.

32. Cortez, "Of Exterior and Exception," 124–50.

33. Gómez-Peña, *New World Border*, 11.

34. San Pedro Creek Preliminary Engineering Report, 31.

35. Mathis, "Survey."

36. San Pedro Creek Subcommittee Meeting Notes. Some of these comments appear to be references to the fact that Muñoz and Company also helped design the Six Flags Theme Park on the Northwest Side of San Antonio.

37. Rivard and Dimmick, "San Pedro Creek Project."

38. Rivard and Dimmick, "San Pedro Creek Project."

39. San Pedro Creek Subcommittee Meeting Notes.

40. San Pedro Creek Subcommittee Meeting Notes.

41. Rivard and Dimmick, "San Pedro Creek Project."

42. Rivard and Dimmick, "San Pedro Creek Project."

43. Vinson, "Elizondo Defends San Pedro Creek Design."

44. Rivard and Dimmick, "San Pedro Creek Project."

45. Rivard and Dimmick, "San Pedro Creek Project."

46. San Pedro Creek Subcommittee Meeting Notes, December.

47. San Pedro Creek Subcommittee Meeting Notes, December.

48. Rivard, "San Pedro Creek Design Changes."

49. Bedoya, "Spatial Justice"; Mignolo, *Local Histories/Global Designs*.

50. Peggy O'Hare, "Soapworks Residents, Activists Demand City Intervention," *San Antonio Express News*, April 12, 2018, www.expressnews.com/news/local.

51. Darcy Sprague, "Like San Pedro Creek, These Apartments Are Being Renovated —and Its Residents Are Worried," *Folo Media*, January 19, 2018, www.folomedia.org.

52. Cortez, "Vecinos de Mission Trails."

53. Cortez, "Vecinos de Mission Trails," 3.

54. Cortez, "Vecinos de Mission Trails," 16.

55. Purcell, "To Inhabit Well," 560–74; Cortez, "No Nos Moveran."

56. Cortez, "No Nos Moveran," 92.

57. Reynolds, *Geographies of Writing*; Muckelbauer, *Future of Invention*, 122.

58. Rickert, *Ambient Rhetoric*, 41.

59. Cortez, "No Nos Moveran," 92; Tsing, *Mushroom at the End of the World*.

60. Boyle, *Rhetoric as a Posthuman Practice*, 146–47.

61. Dimmick, "Housing Policy Task Force."

62. Dimmick, "Housing Policy Task Force."

63. Pelling, *Adaptation to Climate Change*.

64. Rice, *Distant Publics*.

65. Latour, *Inquiry into Modes of Existence*; Haraway, *Staying with the Trouble*.

66. Wallace-Wells, *Uninhabitable Earth*.

Chapter Five

1. The inaugural Rhetoric Society of America (RSA) project on power, place, and publics used a campus master plan to reconsider placed-based rhetorical theory in very similar ways. I'd like to thank John Ackerman for his notion of city plans as long unfulfilled desires.

2. Rosales, *Illusion of Inclusion*.

3. Cintrón, "Abandon All Hope," 189–212.

4. Edwards, "Digital Rhetoric on a Damaged Planet," 59–72.

5. Haraway, *Staying with the Trouble*.

6. City of San Antonio: Office of Equity.

7. SA Climate Ready.

8. Cantu et al., "An Open Letter." See also "From Climate Strike."

9. Wallace-Wells, *Uninhabitable Earth*, 9.

10. Wallace-Wells, *Uninhabitable Earth*, 14.

11. Wallace-Wells, *Uninhabitable Earth*, 11.

12. Wallace-Wells, *Uninhabitable Earth*.

13. Cintrón, "Abandon All Hope," 205.

14. Cintrón, "Abandon All Hope," 208.

15. Tsing, *Mushroom at the End of the World*.

16. Ceccarelli, "Biocolonialism and Human Genomics Research."

17. Paroske, "Deliberating International Science Policy Controversies," 148–70.

18. Happe, *Material Gene.*

19. Druschke, "Trophic Future for Rhetorical Ecologies."

20. Harding, "One Planet, Many Sciences," 41.

21. García and Baca, *Rhetorics Elsewhere and Otherwise*; Haas and Frost, "Toward an Apparent Decolonial Feminist Rhetoric of Risk."

22. García and Baca, *Rhetorics Elsewhere and Otherwise*, 8–9.

23. Chávez, "Beyond Inclusion," 162–72.

24. García and Baca, *Rhetorics Elsewhere and Otherwise*, 24.

25. Michael Shear, Zolan Kanno-Youngs, and Ana Swanson, "Trump Says No Deal with Mexico Is Reached as Border Arrests Surge," *New York Times*, June 5, 2019.

26. Nicholas Kristof, "Food Doesn't Grow Here Anymore. That's Why I Would Send My Son North," *New York Times*, June 5, 2019, www.nytimes.com.

27. Oliver Milman, Emily Holden, and David Argen, "The Unseen Driver behind the Migrant Caravan: Climate Change," *The Guardian*, October 30, 2018, www.theguardian.com.

28. Milman, Holden, and Argen, "Unseen Driver."

29. Milman, Holden, and Argen, "Unseen Driver."

30. Coral Davenport, "Major Climate Report Describes a Strong Risk of Crisis as Early as 2040," *New York Times*, October 7, 2018, www.nytimes.com.

31. SA Climate Ready, 46.

32. Kendra Pierre-Louis, "Want to Escape Global Warming? These Cities Promise Cool Relief," *New York Times*, April 15, 2019, www.nytimes.com; Alyson Krueger, "Climate Change Insurance: Buy Land Somewhere Else," *New York Times*, November 30, 2018, www.nytimes.com.

33. Graham and Walsh, "There's No Such Thing as a Scientific Controversy," 192–206.

34. Norgaard, *Living in Denial.*

35. Nafeez Ahmed, "U.S. Military Could Collapse within 20 Years Due to Climate Change, Report Commissioned by Pentagon Says," *Vice*, October 24, 2019, www.vice.com.

36. Casas, "To See and to Show."

37. SA Climate Ready, 42.

38. Cortez, *Luz at Midnight.*

39. Žižek, "Slavoj Žižek on the Limits of Local Politics."

40. Rai, *Democracy's Lot*, 201–2.

41. Rai, *Democracy's Lot*, 204.

42. The polar vortex event that Texas endured during the week of February 15, 2021, is another example of historical extremes becoming more frequent. Learning from these histories in order to prepare for future climate breakdown events seems necessary. The material difference in learning from and living with these histories is exemplified by El Paso's resilience during this event versus most of the rest of Texas.

43. Ahmed, "U.S. Military Could Collapse within 20 Years."

44. Endres et al., "Communicating Energy."

45. Endres et al., "Communicating Energy," 430–35.

46. Meitzen, "Water Symposium."

47. SA Climate Ready, 46.

48. Christopher Flavelle, "Even as Floods Worsen with Climate Change, Fewer People Insure against Disaster," *New York Times*, June 8, 2019, www.nytimes.com.

49. Goodell, *Water Will Come*, 269.

50. Meitzen, "Water Symposium."

51. Rai, *Democracy's Lot*, 6.

52. Tsing, *Mushroom at the End of the World*, 161.

CODA

1. Sagarin and Pauchard, *Observation and Ecology.*

2. Kimmerer, *Braiding Sweetgrass.*

3. Anzaldúa, *Light in the Dark/Luz En Lo Oscuro.*

Bibliography

Abeles, Oren, Jordynn Jack, and Sarah Singer. "Resilient Turns: Epistrophe, Incrementum, Metonymy." *Project on Rhetoric of Inquiry* 15, no. 1 (2020). https://ir.uiowa.edu /poroi.

Ackerman, John. "The Space for Rhetoric in Everyday Life." In *Towards a Rhetoric of Everyday Life: New Directions in Research on Writing, Text, and Discourse*, edited by Martin Nystrand and John Duffy, 84–117. Madison: University of Wisconsin Press, 2003.

Alcoff, Linda Martín. "The Problem of Speaking for Others." *Cultural Critique* 20 (1991): 24–27.

Alinsky, Saul. *Rules for Radicals: A Pragmatic Primer for Realistic Radicals.* New York: Random House, 1971.

Allen, Danielle. *Talking to Strangers: Anxieties of Citizenship since Brown v. Board of Education.* Chicago: University of Chicago Press, 2009.

Álvarez, Lina, and Brendan Coolsaet. "Decolonizing Environmental Justice Studies: A Latin American Perspective." *Capitalism Nature Socialism* (2018). https://doi-org .libdata.lib.ua.edu/10.1080/10455752.2018.1558272.

Anguiano, Claudia, Tema Milstein, Iliana De Larkin, Yea-Wen Chen, and Jennifer Sandoval. "Connecting Community Voices: Using a Latino/a Critical Race Theory Lens on Environmental Justice Advocacy." *Journal of International and Intercultural Communication* 5, no. 2 (2012): 124–43.

Anzaldúa, Gloria. *Borderlands/La Frontera: The New Mestiza.* 3rd ed. San Francisco: Aunt Lute, 1987.

———. *Luz en lo Oscuro/Light in the Dark: Rewriting Identity, Spirituality, Reality.* Durham, NC: Duke University Press, 2018.

Anzaldúa, Gloria, and AnaLouise Keating, eds. *This Bridge We Call Home: Radical Visions for Transformation.* New York: Routledge, 2002.

Asen, Robert. "A Discourse Theory of Citizenship." *Quarterly Journal of Speech* 90, no. 2 (2004): 189–211.

"Atlas 14, Volume 11." National Oceanic and Atmospheric Administration. Last modified April 21, 2017. https://hdsc.nws.noaa.gov.

"Authenticity." *Oxford English Dictionary.* Accessed September 10, 2019. www.oed.com. libweb.lib.utsa.edu.

Awan, Nishat. "Introduction to Border Topologies." *GeoHumanities* 2, no. 2 (2016): 279–83.

Baca, Damián. *Mestiz@ Scripts, Digital Migrations, and the Territories of Writing*. New York: Palgrave Macmillan, 2008.

Baca, Damián, and Victor Villanueva, eds. *Rhetorics of the Americas: 3114 BCE to 2012 CE*. New York: Palgrave Macmillan, 2010.

Barr, Juliann. "Beyond Their Control: Spaniards in Texas." In *Choice, Persuasion, and Coercion: Social Control on Spain's North American Frontiers*, edited by Jesús F. de la Teja and Ross Frank, 149–77. Albuquerque: University of New Mexico Press, 2005.

Bean, Hamilton, Lisa Keränen, and Margaret Durfy. "'This Is London': Cosmopolitan Nationalism and the Discourse of Resilience in the Case of the 7/7 Terrorist Attacks." *Rhetoric and Public Affairs* 14, no. 3 (2011): 427–64.

Bedoya, Roberto. "Spatial Justice: Rasquachification, Race, and the City." September 15, 2014. http://creativetimereports.org.

Bendell, Jem. "Six Months of the Deep Adaptation Forum." September 9, 2019. https://jembendell.com.

Berriozábal, María Antonietta. *María, Daughter of Immigrants*. San Antonio: Wings Press, 2012.

Black Congressional Caucus. *African Americans and Climate Change: An Unequal Burden*. Accessed on September 6, 2019. http://sustainablecommunitydevelopmentgroup.org.

Booher, Amanda, and Julie Jung, eds. *Feminist Rhetorical Science Studies: Human Bodies, Posthuman Worlds*. Carbondale: Southern Illinois University Press, 2018.

Boyle, Casey. *Rhetoric as a Posthuman Practice*. Columbus: Ohio State University Press, 2018.

Bretcher, Jeremy. "Making the Green New Deal Work for Workers." *In These Times*, April 22, 2019. http://inthesetimes.com.

Brischetto, R., C. Cottrell, and R. M. Stevens. "Conflict and Change in the Political Culture of San Antonio in the 1970s." In *The Politics of San Antonio: Community, Progress, and Power*, 75–94, edited by David R. Johnson, John Booth, and Richard Harris. Lincoln: University of Nebraska Press, 1983.

Brouwer, David, and Robert Asen. *Public Modalities: Rhetoric, Culture, Media, and the Shape of Public Life*. Tuscaloosa: University of Alabama Press, 2010.

Buitron, Richard. *The Quest for Tejano Identity in San Antonio, Texas, 1913–2000*. New York: Routledge, 2004.

Bullard, Robert. *Dumping in Dixie: Race, Class, and Environmental Quality*. Boulder, CO: Westview Press, 1987.

Burke, Kenneth. *A Grammar of Motives*. Berkeley: University of California Press, 1969.

———. *The Philosophy of Literary Form: Studies in Symbolic Action*. Baton Rouge: Louisiana State University Press, 1941.

———. *A Rhetoric of Motives*. Berkeley: University of California Press, 1962.

Cadena, Marisol de la. *Earth Beings: Ecologies of Practice across Andean Worlds*. Durham, NC: Duke University Press, 2015.

Cadena, Marisol de la, and Mario Blaser, eds. *A World of Many Worlds*. Durham, NC: Duke University Press, 2018.

Cantu, Adelita G., Alfred Montoya, Diana Lopez, Elizabeth Montgomery, Graciela

Sanchez, Jessica O. Guerrero, Leslie Provence, and Tim Barr. "An Open Letter to Mayor Ron Nirenberg and the Members of the San Antonio City Council," September 20, 2020.

Casas, Ruben. "To See and to Show: Maps as (Neo)colonial Technologies and the Potential in Countervisuality." In *In/visibility, Mobility, and Making Do in Contemporary Latina/o Migrant Rhetorics*. PhD diss., University of Wisconsin–Madison, 2019.

Ceccarelli, Leah. "Biocolonialism and Human Genomics Research: The Frontier Mapping Expedition of Francis Collins." In *On the Frontier of Science: An American Rhetoric of Exploration and Exploitation*, 91–110. East Lansing: Michigan State University Press, 2013.

Certeau, Michel de. *The Practice of Everyday Life*. Berkeley: University of California Press, 1984.

Charland, Maurice. "Constitutive Rhetoric: The Case of the Peuple Québécois." *Quarterly Journal of Speech* 73, no. 2 (1987): 133–50.

Chávez, Karma. "Beyond Inclusion: Rethinking Rhetoric's Historical Narrative." *Quarterly Journal of Speech* 101, no. 1 (2015): 162–72.

———. *Queer Migration Politics: Activist Rhetoric and Coalitional Possibilities*. Urbana: University of Illinois Press, 2013.

Cintrón, Ralph. "Abandon All Hope Ye Who Enter Here: Democracy and Climate Change." *Works and Days: The 40th Anniversary Retrospective: Capitalism, Climate Change, and Rhetoric* 36 (2018–19): 189–212.

———. *Angel's Town: Chero Ways, Gang Life, and Rhetorics of the Everyday*. Boston: Beacon Press, 1997.

Cisneros, Josue David. *The Border Crossed Us: Rhetorics of Borders, Citizenship, and Latina/o Identity*. Tuscaloosa: University of Alabama Press, 2014.

City of San Antonio: Office of Equity. Accessed on September 9, 2019. www.sanantonio.gov.

"Climate Mayors." Accessed January 5, 2017. http://climatemayors.org.

Condit, Celeste. *The Meanings of the Gene: Public Debates about Human Heredity*. Madison: University of Wisconsin Press, 1999.

Cortez, José. "Of Exterior and Exception: Latin American Rhetoric, Subalternity, and the Politics of Cultural Difference." *Philosophy and Rhetoric* 51, no. 21 (2018): 124–50.

Cortez, Marisol. *Luz at Midnight*. San Antonio: Flowersong Press, 2020.

———. "No Nos Moveran: Embodying *Buen Vivir* in the Case of Mission Trails Mobile Home Community." In *Community as the Material Basis of Citizenship: The Unfinished Story of American Democracy*, edited by Rodolfo Rosales, 75–100. New York: Routledge, 2018.

———. "Vecinos de Mission Trails, Preliminary Findings and Analysis from a Case Study on the Displacement of Mission Trails Mobile Home Community." *Report Presented to the City of San Antonio's Housing Commission on Protecting and Preserving Diverse and Dynamic Neighborhoods*. San Antonio, Texas, March 20, 2016. https://vecinosdemissiontrails.

Cushman, Ellen, Rachel Jackson, Annie Laurie Nichols, Courtney Rivard, Amanda Moulder, Chelsea Murdock, David M. Grant, and Heather Brook Adams.

"Decolonizing Projects: Creating Pluriversal Possibilities in Rhetoric." *Rhetoric Review* 38, no. 1 (2019): 1–22.

Danisch, Robert. *Pragmatism, Democracy, and the Necessity of Rhetoric.* Columbia: University of South Carolina Press, 2007.

DeChaine, Robert, ed. *Border Rhetorics: Citizenship and Identity on the US-Mexico Frontier.* Tuscaloosa: University of Alabama Press, 2012.

De Los Santos, René Agustín. "'The Future of Our History': Rhetorics of Transformation and Power in Plutarco Elías Calles' 1928 Informe." *Rhetoric Society Quarterly* 45 no. 3 (2015): 199–211.

Dimmick, Iris. "Housing Policy Task Force Recommends Sweeping Changes to Alleviate Market, Policy Shortfalls." *Rivard Report.* June 20, 2018. https://therivardreport .com.

Donecker, Frances. "The San Antonio Express News." *Handbook of Texas Online.* Accessed March 22, 2017. www.tshaonline.org.

———. "San Antonio Light." *Handbook of Texas Online.* Accessed March 22, 2017. www .tshaonline.org.

Dove, Michael, and Carol Carpenter, eds. *Environmental Anthropology: A Historical Reader.* New York: Wiley-Blackwell, 2007.

Druschke, Caroline G. "A Trophic Future for Rhetorical Ecologies." *enculturation: A Journal of Rhetoric, Writing, and Culture* (2019). http://enculturation.net.

Druschke, Caroline G., and Bridie McGreavy. "Why Rhetoric Matters for Ecology." *Frontiers in Ecology and the Environment* 14, no. 1 (2016): 46–52.

Druschke, Caroline G., and Candice Rai. "Making Worlds with Cyborg Fish." In *Tracing Rhetoric and Material Life: Ecological Approaches,* edited by Bridie McGreavy, Justine Wells, George F. McHendry Jr., and Samantha Senda-Cook, 197–222. Cham, Switzerland: Palgrave, 2018.

Druschke, Caroline G., Nedra Reynolds, Jennifer Morton-Aiken, Ingrid E. Lofgren, Nancy E. Karraker, and Scott R. McWilliams. "Better Science through Rhetoric: A New Model and Pilot Program for Training Graduate Student Science Writers." *Technical Communication Quarterly* 27, no. 2 (2018): 175–90.

Edbauer, Jenny. "Unframing Models of Public Distribution: From Rhetorical Situations to Rhetorical Ecologies." *Rhetoric Society of America* 35, no. 4 (2005): 5–24.

Edwards, Dustin W. "Digital Rhetoric on a Damaged Planet: Storying Digital Damage as Inventive Response to the Anthropocene." *Rhetoric Review* 39, no. 1 (2020): 59–72.

Endres, Danielle, Brian Cozen, Joshua Trey Barnett, Megan O'Byrne, and Tarla Rai Peterson. "Communicating Energy in a Climate (of) Crisis." In *Communication Yearbook 40,* edited by Ellsia L. Cohen, 419–47. New York: Routledge, 2016.

Escobar, Arturo. *Designs for the Pluriverse: Radical Interdependence, Autonomy, and the Making of Worlds.* Durham, NC: Duke University Press, 2017.

Eubanks, Phillip. *The Troubled Rhetoric and Communication of Climate Change.* New York: Routledge, 2015.

Fahnestock, Jeanne. *Rhetorical Figures in Science.* New York: Oxford University Press, 2002.

Fisher, Lewis. *American Venice: The Epic Story of San Antonio's River.* San Antonio: Maverick Publishing, 2015.

Flood Relief Committee. *Memorandum to the San Antonio Board of Engineers.* University of Texas at San Antonio Libraries Special Collections, 1921.

Flynn, Elizabeth, Patricia Sotirin, and Ann Brady, eds. *Feminist Rhetorical Resilience.* Logan: Utah State University Press, 2012.

Foucault, Michel. *The Order of Things: An Archaeology of the Human Sciences.* New York: Vintage, 1994.

Fourth National Climate Assessment, vols. 1 and 2. November 2018. www.globalchange .gov.

"From Climate Strike to Climate Thunder: Unearthing Environmental Resistance in South Texas." *Urban-15: Hidden Histories.* September 29, 2019. https://urban15.org.

Frost, Samantha. *Biocultural Creatures: Toward a New Theory of the Human.* Durham, NC: Duke University Press, 2018.

García, Romeo, and Damián Baca. *Rhetorics Elsewhere and Otherwise: Contested Modernities, Decolonial Visions.* New York: National Council of Teachers of English, 2019.

Gibbons, Brendon. "NOAA Study Shows Heavy Rains Have Grown More Frequent in Texas." *San Antonio Report.* November 11, 2018. https://therivardreport.com.

Gómez-Peña, Guillermo. *The New World Border: Prophecies, Poems, and Loqueras for the End of the Century.* San Francisco: City Lights, 1996.

Goodell, Jeff. *The Water Will Come: Rising Seas, Sinking Cities, and the Remaking of the Civilized World.* New York: Little, Brown, 2017.

Goodwin, Jean, ed. *Between Scientists and Citizens: Proceedings of a Conference at Iowa State University, June 1–2, 2012.* CreateSpace Independent Publishing Platform, 2012.

Goodwin, Jean, Michael F. Dahlstrom, Mari Kemis, Clark Wolf, and Christine Hutchison. "Rhetorical Resources for Teaching Responsible Communication of Science." *Project on Rhetoric of Inquiry (POROI)* 10, no. 1 (2014). https://ir.uiowa.edu/poroi/.

Graham, Scott S. "Object-Oriented Ontology's Binary Duplication and the Promise of Thing-Oriented Ontologies." In *Rhetoric, through Everyday Things*, edited by Scot Barnett and Casey Boyle, 108–24. Tuscaloosa: University of Alabama Press, 2016.

———. *The Politics of Pain Medicine: A Rhetorical-Ontological Inquiry.* Chicago: University of Chicago Press, 2015.

———. "The Promise and Peril of Scientific Science Studies." *Theory and Event* 21, no. 2 (2018): 529–53.

Graham, Scott S., and Lynda Walsh. "There's No Such Thing as a Scientific Controversy." *Technical Communication Quarterly* 28, no. 3 (2019): 192–206.

Gries, Laurie. *Still Life with Rhetoric: A New Materialist Approach for Visual Rhetorics.* Logan: Utah State University Press, 2015.

Haas, Angela, and Erin Frost. "Toward an Apparent Decolonial Feminist Rhetoric of Risk." In *Topic-Driven Environmental Rhetoric*, edited by Derek Ross, 168–86. New York: Routledge, 2017.

Haggard, J. V. "Spain's Indian Policy in Texas." *Southwestern Historical Quarterly* 46 (1942): 76–77.

Happe, Kelly. *The Material Gene: Gender, Race, and Heredity after the Human Genome Project.* New York: New York University Press, 2013.

Haraway, Donna. "Awash in Urine: DES and Premarin in Multispecies

Response-ability." *Women's Studies Quarterly* 40, no. 1/2 (Spring/Summer 2012): 301–16.

———. "The Promises of Monsters: A Regenerative Politics for Inappropriate/d Others." *The Donna Haraway Reader*. New York: Routledge, 2004.

———. "Situated Knowledges: The Science Question in Feminism and the Privilege of Partial Perspective." *Feminist Studies* 14, no. 3 (1988): 575–99.

———. *Staying with the Trouble: Making Kin in the Chthulucene*. Durham, NC: Duke University Press, 2015.

Harding, Sandra. "Latin American Decolonial Social Studies of Scientific Knowledge: Alliances and Tensions." *Science, Technology, and Human Values* 41, no. 6 (2016): 1063–87.

———. "One Planet, Many Sciences." In *Constructing the Pluriverse: The Geopolitics of Knowledge*, edited by Bernd Reiter, 39–62. Durham, NC: Duke University Press, 2018.

———. *Sciences from Below: Feminisms, Postcolonialites, and Modernities*. Durham, NC: Duke University Press, 2008.

———. "State of the Field: Latin American Decolonial Philosophies of Science." *Studies in History and Philosophy of Science* 78 (2019): 48–63.

Harris, Randy Allen. "Dementia, Rhetorical Schemes, and Cognitive Resilience." *Project on Rhetoric of Inquiry* 15, no. 1 (2020).

Hawhee, Debra, and Christa Olson. "Pan-Historiography: The Challenges of Writing History across Time and Space." In *Theorizing Histories of Rhetoric*, edited by Michelle Ballif, 90–105. Carbondale: Southern Illinois University Press, 2013.

Hawk, Byron. *Resounding the Rhetorical: Composition as Quasi-Object*. Pittsburgh: University of Pittsburgh Press, 2018.

Hayhoe, Katherine. "Climate Trends in San Antonio and an Overview of Climate Projections for the South Central Region." *ATMOS Research and Consulting: San Antonio Sustainability Plan, Climate Report: Appendix C*, 2015. www.sasustainabilityplan .com.

Hernandez-Ehrisman, Laura. *Inventing the Fiesta City: Heritage and Carnival in San Antonio*. Albuquerque: University of New Mexico Press, 2008.

Herndl, Carl, and Stuart Brown, eds. *Green Culture: Environmental Rhetoric in Contemporary America*. Madison: University of Wisconsin Press, 1997.

Herndl, Carl, and Lauren Cutlip. "'How Can We Act?' A Praxiographical Program for the Rhetoric of Technology, Science, and Medicine." *Project on Rhetoric of Inquiry* 9, no. 1 (2013).

Herndl, Carl, and S. Scott Graham. "Getting Over Incommensurability: Latour, New Materialisms, and the Rhetoric of Diplomacy." In *Thinking with Bruno Latour in Rhetoric and Composition*, edited by Paul Lynch and Nathaniel Rivers, 40–58. Carbondale: Southern Illinois University Press, 2015.

Herndl, Carl, Sarah Beth Hopton, Lauren Cutlip, Elena Yu Polush, Rick Cruse, and Mack Shelley. "What's a Farm? The Languages of Space and Place." In *Field Rhetoric: Ethnography, Ecology, and Engagement in the Places of Persuasion*, edited by Candice Rai and Caroline Gottschalk Druschke, 61–94. Tuscaloosa: University of Alabama Press, 2018.

Herndl, Carl G., and Adela C. Licona. "Shifting Agency: Agency, *Kairos*, and the Possibilities of Social Action." In *Communicative Practices in Workplaces and the Professions: Cultural Perspectives on the Regulation of Discourse and Organizations*, edited by Mark Zachry and Charlotte Thralls, 133–54. New York: Routledge, Baywood Press, 2007.

Hesford, Wendy, Adela Licona, and Christa Teston, eds. *Precarious Rhetorics*. Columbus: Ohio State University Press, 2018.

Housing Rights and Climate Justice. *Climate Action SA: Healthy Families, Strong Communities, and a Sustainable Future*. Accessed November 15, 2018. https://youtu.be /jl7tW2KNNPE.

Hulme, Mike. *Weathered: Cultures of Climate*. New York: Sage, 2016.

Intergovernmental Panel on Climate Change (IPCC). "Summary for Policy Makers," 1995. Accessed May 17, 2020. https://library.harvard.edu.

———. "Summary for Policy Makers," 2018. Accessed May 17, 2020. www.ipcc.ch.

Jack, Jordynn, *Autism and Gender: From Refrigerator Mothers to Computer Geeks*. Urbana: University of Illinois Press, 2014.

———. *Science on the Home Front: American Women Scientists in World War II*. Urbana: University of Illinois Press, 2009.

Jackson, Rachel. "Decolonizing Place and Race: Racial Resentments, Local Histories, and Transrhetorical Analysis." In "Rhetoric, Race, and Resentment: Whiteness and the New Days for Rage," by Meta G. Carstarphen, Kathellen E. Welch, Wendy K. Z. Anderson, Davis W. Houck, Mark L. McPhail, David A. Frank, Rachel C. Jackson, James Alexander McVey, Christopher J. Gilbert, Patricia G. Davis, and Lisa M. Corrigan. *Rhetoric Review* 36, no. 4 (2017): 292–302.

Jarboe, Jan. "Building a Movement: Mexican American Struggles for Municipal Services." *Civil Rights Digest* (1977): 38–46.

Johnson, Jenell. *American Lobotomy: A Rhetorical History*. Ann Arbor: University of Michigan Press, 2014.

Keeling, Diane M., and Jennifer C. Prairie. "Trophic and Tropic Dynamics: An Ecological Perspective of Tropes." In *Tracing Rhetoric and Material Life: Ecological Approaches*, edited by Bridie McGreavy, Justine Wells, George F. McHendry Jr., and Samantha Senda-Cook, 39–58. Cham, Switzerland: Palgrave, 2018.

Kelly, Casey. "Détournement, Decolonization, and the American Indian Occupation of Alcatraz Island (1969–1971)." *Rhetoric and Public Affairs* 44, no. 2 (2014): 168–90.

Keränen, Lisa. *Scientific Characters: Rhetoric, Politics, and Trust in Breast Cancer Research*. Tuscaloosa: University of Alabama Press, 2010.

Killingsworth, Jimmie, and Jacqueline Palmer. *Ecospeak: Rhetoric and Environmental Politics in America*. Carbondale: Southern Illinois University Press, 1992.

Kimmerer, Robin Wall. *Braiding Sweetgrass: Indigenous Wisdom, Scientific Knowledge, and the Teaching of Plants*. Minneapolis: Milkweed, 2013.

Klein, Naomi. *This Changes Everything: Capitalism vs. the Climate*. New York: Simon and Schuster, 2015.

Klinenberg, Eric. "Want to Survive Climate Change? You'll Need a Good Community." *Wired*, October 25, 2016. www.wired.com.

Latour, Bruno. *An Inquiry into Modes of Existence: An Anthropology of the Moderns.* Translated by Catherine Porter. Cambridge, MA: Harvard University Press, 2015.

———. *Science in Action: How to Follow Scientists and Engineers through Society.* Cambridge, MA: Harvard University Press, 1988.

———. *We Have Never Been Modern.* Translated by Catherine Porter. Cambridge, MA: Harvard University Press, 1993.

Lawhn, J. L. "María Luisa Garza: Novelist of *el México de afuera.*" In *Double Crossings/EntreCruzamientos,* edited by M. Flores and C. Von Son, 83–96. New York: Ediciones Nuevo Espacio, 2001.

Libreria de Quiroga, ed. *La Tragedia de la Inundacion de San Antonio: Un Recuerdo de la Terrible Catástrofe.* 2nd ed. San Antonio: Libreria de Quiroga [Qiroga], 1921.

Licona, Adela. *Zines in Third Space: Radical Cooperation and Borderlands Rhetoric.* Albany: State University of New York Press, 2012.

Licona, Adela, and Karma Chávez. "Relational Literacies and Their Coalitional Possibilities." *Peitho Journal* 18, no. 1 (2015): 96–107.

Liévanos, Raoul. "A Minority Perspective Is Limited: Environmental Privilege and Surface Water Hazards in an Impaired Estuary." *Association of American Geographers,* April 18, 2010.

Lind, Dara. "The '500-Year' Flood Explained: Why Houston Was So Unprepared for Hurricane Harvey." Last updated April 28, 2017. *Vox.* www.vox.com.

Lipsitz, George. *How Racism Takes Place.* Philadelphia: Temple University Press, 2011.

Logan, John, and Harvey Molotch. *Urban Fortunes: The Political Economy of Place.* Berkeley: University of California Press, 1987.

Lugones, María. *Pilgrimages/Peregrinajes: Theorizing Coalition against Multiple Oppressions.* Lanham, MD: Rowman and Littlefield, 2005.

Lyon, Arabella. *Deliberative Acts: Democracy, Rhetoric, and Rights.* University Park: Pennsylvania State University Press, 2013.

Martinez, Monica Muñoz. *The Injustice Never Leaves You: Anti-Mexican Violence in Texas.* Cambridge, MA: Harvard University Press, 2018.

Mathis, Dan. "Survey: Public Supports San Pedro Creek Improvements." *Rivard Report.* Accessed September 9, 2019. http://therivardreport.com.

McGreavy, Bridie. "Resilience as Discourse." *Environmental Communication* 10, no. 1 (2016): 104–21.

McGreavy, Bridie, Justine Wells, Guy McHendry, and Samantha Senda-Cook, eds. *Tracing Rhetoric and Material Life: Ecological Approaches.* New York: Palgrave Macmillan, 2017.

McKinnon, Sara, Rob Asen, Karma Chávez, Robert Glen Howard, eds. *text + FIELD: Innovations in Rhetorical Method.* University Park: Pennsylvania State University Press, 2016.

Medina, Cruz. *Reclaiming Poch@ Pop: Examining the Rhetoric of Cultural Deficiency.* New York: Palgrave Macmillan, 2015.

Mehlenbacher, Ashley R. *Science Communication Online: Engaging Experts and Publics on the Internet.* Columbus: Ohio State University Press, 2019.

Meitzen, Kimberly. "Water Symposium Addresses the Future of Flooding in Texas."

Rivard Report. November 24, 2018. https://therivardreport.com.

Mejía, Jaime Armin. "Tejano Arts of the U.S.-Mexico Contact Zone." *JAC* 18, no. 1 (1998): 123–35.

Mignolo, Walter. *The Darker Side of Western Modernity: Global Futures, Decolonial Options.* Durham, NC: Duke University Press, 2011.

———. "Foreword: On Pluriversality and Multipolarity." In *Constructing the Pluriverse: The Geopolitics of Knowledge,* edited by Bernd Reiter, ix–xvi. Durham, NC: Duke University Press, 2018.

———. *Local Histories/Global Designs: Coloniality, Subaltern Knowledges, and Border Thinking.* Princeton, NJ: Princeton University Press, 2012.

Mignolo, Walter, and Catherine Porter. *On Decoloniality: Concepts, Analytics, Praxis.* Durham, NC: Duke University Press, 2018.

Miller, Carolyn. "Genre as Social Action." *Quarterly Journal of Speech* 70 (1984): 151–76.

Miller, Char. *San Antonio: A Tricentennial History.* San Antonio: City of San Antonio, 2018.

———. "Streetscape Environmentalism: Floods, Social Justice, and Political Power in San Antonio, 1921–1974." *Southwestern Historical Quarterly* 118, no. 2 (2004): 158–77.

Montejano, David. *Anglos and Mexicans in the Making of Texas, 1836–1986.* Austin: University of Texas Press, 1987.

———. *Quixote's Soldiers: A Local History of the Chicano Movement, 1966–1981.* Austin: University of Texas Press, 2010.

Moraga, Cherríe, and Gloria Anzaldúa, eds. *This Bridge Called My Back: Writings by Radical Women of Color.* Albany: State University of New York Press, 1981.

Muckelbauer, John. *The Future of Invention: Rhetoric, Postmodernism, and the Problem of Change.* Albany: State University of New York Press, 2004.

———. "Implicit Paradigms of Rhetoric: Aristotelian, Cultural, Heliotrophic." In *Rhetoric, through Everyday Things,* edited by Scot Barnett and Casey Boyle, 30–41. Tuscaloosa: University of Alabama Press, 2016.

Muñoz, Henry. *Muñoz.* Accessed September 9, 2019. http://munozandcompany.com/.

Norgaard, Kari. *Living in Denial: Climate Change, Emotions, and Everyday Life.* Cambridge, MA: MIT Press, 2011.

Olson, Christa. *Constitutive Visions: Indigeneity and Commonplaces of National Identity in Republican Ecuador.* University Park: Pennsylvania State University Press, 2014.

Olson, Christa, and René Agustín De los Santos. "Expanding the Idea of América." *Rhetoric Society Quarterly* 45, no. 3 (2015): 193–98.

Onís, Kathleen de. "'Looking Both Ways': Metaphor and the Rhetorical Alignment of Intersectional Climate Justice and Reproductive Justice Concerns." *Environmental Communication* 6, no. 3 (2012): 308–27.

Orozco, Cynthia. *No Mexicans, Women, or Dogs Allowed: The Rise of the Mexican American Civil Rights Movement.* Austin: University of Texas Press, 2009.

Paliewicz, Nicholas. "Bent but Not Broken: Remembering Vulnerability and Resiliency at the National September 11 Memorial Museum." *Southern Communication Journal* 82, no. 1 (2017): 1–14.

Paroske, Marcus. "Deliberating International Science Policy Controversies: Uncertainty and AIDS in South Africa." *Quarterly Journal of Speech* 95, no. 2 (2009): 148–70.

Pelling, Mark. *Adaptation to Climate Change: From Resilience to Transformation*. New York: Routledge, 2011.

Peña, Devon. *Mexican Americans and the Environment: Tierra y Vida*. Tucson: University of Arizona Press, 2015.

Peterson, Tarla Rai. *Sharing the Earth: The Rhetoric of Sustainable Development*. Columbia: University of South Carolina Press, 1997.

Pezzullo, Phaedra. *Toxic Tourism: Rhetorics of Pollution, Travel, and Environmental Justice*. Tuscaloosa: University of Alabama Press, 2007.

Plotkin, Sydney. "Democratic Change in the Urban Political Economy: San Antonio's Edwards Aquifer Controversy." In *The Politics of San Antonio: Community, Progress, and Power*, edited by D. R. Johnson, J. Booth, and R. Harris, 157–74. Lincoln: University of Nebraska Press, 1983.

Powell Malea, Daisy Levy, Andrea Riley-Mukavetz, Marilee Brooks-Gillies, Maria Novotny, and Jennifer Fisch-Ferguson. "Our Story Begins Here: Constellating Cultural Rhetorics." *Enculturation: A Journal of Rhetoric, Writing, and Culture* (October 2014): http://enculturation.net.

Pulido, Laura. "Rethinking Environmental Racism: White Privilege and Urban Development in Southern California." *Annals of the Association of American Geographers, Association of American Geographers* 90 (2000): 12–40.

Purcell, Mark. "To Inhabit Well: Counterhegemonic Movements and the Right to the City." *Urban Geography* 34, no. 4 (2013): 560–74.

Quijano, Aníbal. "Coloniality and Modernity/Rationality." *Cultural Studies* 21, no. 2–3 (2007): 168–78.

Rai, Candice. *Democracy's Lot: Rhetoric, Publics, and the Places of Persuasion*. Tuscaloosa: University of Alabama Press, 2016.

Rai, Candice, and Caroline Gottschalk Druschke, eds. *Field Rhetoric: Ethnography, Ecology, and Engagement in the Places of Persuasion*. Tuscaloosa: University of Alabama Press, 2018.

Ramírez, Christina. *Occupying Our Space: The Mestiza Rhetorics of Mexican Women Journalists and Activists, 1875–1942*. Tucson: University of Arizona Press, 2015.

Reeves, Joshua. "Suspended Identification: Atopos and the Work of Public Memory." *Philosophy and Rhetoric* 46, no. 3 (2013): 306–27.

Reiter, Bernd, ed. *Constructing the Pluriverse: The Geopolitics of Knowledge*. Durham, NC: Duke University Press, 2018.

Reynolds, Nedra. *Geographies of Writing: Inhabiting Places and Encountering Difference*. Carbondale: Southern Illinois University Press, 2004.

Rice, Jenny. *Distant Publics: Development Rhetoric and the Subject of Crisis*. Pittsburgh: University of Pittsburgh Press, 2012.

Rickert, Thomas. *Ambient Rhetoric: Attunements of Rhetorical Being*. Pittsburgh, PA: University of Pittsburgh Press, 2014.

Ríos, Gabriela. "Cultivating Land-Based Literacies and Rhetorics." *Literacy in Composition Studies* 3, no. 1 (2015): 61–70.

Rivard, Robert. "San Pedro Creek Design Changes Praised, Pose Challenges." *San Antonio Report*, December 5, 2015. https://sanantonioreport.org/.

Rivard, Robert, and Iris Dimmick. "The San Pedro Creek Project: Getting It Right." *Rivard Report*, 2015. http://therivardreport.com.

Rivers, Nathaniel. "Deep Ambivalence and Wild Objects: Toward a Strange Environmental Rhetoric." *Rhetoric Society Quarterly* 45, no. 5 (2015): 420–40.

Roberts, David. "Hurricane Katrina Showed What 'Adapting to Climate Change' Looks Like." *Vox*. August 24, 2015. www.vox.com.

Rodríguez, Iokiñe, and Mirna Liz Inturias, "Conflict Transformation in Indigenous Peoples' Territories: Doing Environmental Justice with a 'Decolonial Turn.'" *Development Studies Research* 5, no. 1 (2018).

Rogers, Mary Beth. *Cold Anger: A Story of Faith and Power Politics*. Denton: University of North Texas Press, 1990.

Rosales, Rodolfo. *The Illusion of Inclusion: The Untold Political Story of San Antonio, Texas*. Austin: University of Texas Press, 2000.

Ross, Derek, ed. *Topic-Driven Environmental Rhetoric*. New York: Routledge, 2017.

SA Climate Ready: A Pathway for Climate Action and Adaptation. Accessed on September 9, 2019. https://saclimateready.org/.

Sackey, Donnie. "An Environmental Justice Paradigm for Technical Communication." In *Key Theoretical Frameworks: Teaching Technical Communication in the Twenty-First Century*, edited by Angela Haas and Michelle Eble, 138–62. Logan: Utah State University Press, 2018.

Sagarin, Rafe, and Aníbal Pauchard. *Observation and Ecology: Broadening the Scope of Science to Understand a Complex World*. Washington, DC: Island Press, 2012.

Saldívar, José David. *Border Matters: Remapping American Cultural Studies*. Berkeley: University of California Press, 1997.

San Pedro Creek Preliminary Engineering Report. May 16, 2013. https://spcculturepark.com.

San Pedro Creek Subcommittee Meeting Notes. July/August/September/December, 2018. Accessed September 9, 2020: https://fx8ez2bioiz3t7vbo9jbdd5l-wpengine.netdna-ssl.com/.

Sandler, Ronald, and Phaedra Pezzullo, eds. *Environmental Justice and Environmentalism: The Social Justice Challenge to the Environmental Movement*. Cambridge, MA: MIT Press, 2007.

Sandoval, Chela. *Methodology of the Oppressed*. Minneapolis: University of Minnesota Press, 2000.

———. "New Sciences: Cyborg Feminism and the Methodology of the Oppressed." In *The Cybercultures Reader*, edited by David Bell and Barbara M. Kennedy. New York: Routledge, 2000.

Sandoval, Moises. *Our Legacy: The First Fifty Years*. Washington, DC: League of United Latin American Citizens (LULAC), 1979, 4–7.

Santos, John Phillip. *De Unos Lugares Perdidos: Of a Few Places Lost in Time, Cuentos/Elegies: St. Anthony's Lost and Found*. San Antonio: City of San Antonio Department of Arts and Culture, 2018.

———. "San Antonio Is a City of Metamorphosis." *Texas Monthly*, May 2018. www.texasmonthly.com.

Schlosberg David, and Lisette Collins. "From Environmental to Climate Justice: Climate Change and the Discourse of Environmental Justice." *Wiley Interdisciplinary Reviews: Climate Change* 5, no. 3 (2014).

Schneider, Tsim D., and Lee M. Panich. "Native Agency at the Margins of Empire: Indigenous Landscapes, Spanish Missions, and Contested Histories." In *Indigenous Landscapes and Spanish Missions*, edited by Lee M. Panich and Tsim D. Schneider, 1–23. Tucson: University of Arizona Press, 2014.

Scott, Blake J. "Extending Rhetorical-Cultural Analysis: Transformations of Home HIV Testing." *College English* 65 (2003): 351–53.

———. *Risky Rhetoric: AIDS and the Cultural Practices of HIV Testing.* Carbondale: Southern Illinois University Press, 2003.

Sekul, J. "Communities Organized for Public Service: Citizen Power and Public Policy in San Antonio." In *The Politics of San Antonio: Community, Progress, and Power*, edited by David R. Johnson, John Booth, and Richard Harris, 175–90. Lincoln: University of Nebraska Press, 1983.

Sharif, Hatim. "Climate Projections for the City of San Antonio." SAClimateReady.org. http://saclimateready.org.

Simmons, Michelle. *Participation and Power: Civic Discourse in Environmental Policy Decisions.* Albany: State University of New York Press, 2007.

Sowards, Stacey. "Environmental Justice in International Contexts: Understanding Intersections for Social Justice in the Twenty-First Century." *Environmental Communication* 6, no. 3 (2012): 37–41.

Spinuzzi, Clay. *Tracing Genres through Organizations: A Sociocultural Approach to Information Design.* Cambridge, MA: MIT Press, 2003.

Steinberg, Ted. *Acts of God: The Unnatural History of Natural Disaster in America.* New York: Oxford University Press, 2000.

Stengers, Isabelle. *Another Science Is Possible: A Manifesto for Slow Science.* Medford, MA: Polity Press, 2018.

———. "The Challenge of Ontological Politics." In *A World of Many Worlds*, edited by Marisol de la Cadena and Mario Blaser, 83–111. Durham, NC: Duke University Press, 2018.

———. "The Cosmopolitical Proposal." In *Making Things Public: Atmospheres of Democracy*, edited by Bruno Latour and Peter Weibel, 994–1003. Cambridge, MA: MIT Press, 2005.

———. "Including Nonhumans in Political Theory: Opening Pandora's Box?" In *Political Matter: Technoscience, Democracy, and Public Life*, edited by Bruce Braun and Sarah J. Whatmore, 3–31. Minneapolis: University of Minnesota Press, 2010.

———. "Introductory Notes on an Ecology of Practices." *Cultural Studies Review* 11, no. 1 (2005): 183–96.

———. "It's Matters of Concern All the Way Down." *ctrl-z: New Media Philosophy* 7 (2016).

Stormer, Nathan. "Articulation: A Working Paper on *Taxis.*" *Quarterly Journal of Speech* 90, no. 3 (2004): 257–84.

Stormer, Nathan, and Bridie McGreavy. "Thinking Ecologically about Rhetoric's

Ontology: Capacity, Vulnerability, and Resilience." *Philosophy and Rhetoric* 50, no. 1 (2017): 1–25.

Taub, Ben. "Inequality and Hurricane Harvey." *New Yorker*, September 6, 2017. www.newyorker.com.

Tlostanova, Madina, and Walter Mignolo. "On Pluritopic Hermeneutics, Trans-modern Thinking, and Decolonial Philosophy." *Encounters* 1, no. 1 (2009): 16.

Tsing, Anna. *The Mushroom at the End of the World: On the Possibility of Life in Capitalist Ruins.* Princeton, NJ: Princeton University Press, 2015.

Tuck, Eve, and K. Wayne Yang. "Decolonization Is Not a Metaphor." *Decolonization: Indigeneity, Education, and Society* 1, no. 1 (2012): 1–40.

Vinson, Joan. "Elizondo Defends San Pedro Creek Design." *Rivard Report*, October 25, 2015. http://therivardreport.com.

Walker, Kenneth. "Mapping the Contours of Translation: Visualized Un/certainties in the Ozone Hole Controversy." *Technical Communication Quarterly* 25, no. 2 (2016): 104–20.

———. "Rhetorical Principles on Uncertainty for Transdisciplinary Engagement and Improved Climate Risk Communication." *Project on Rhetoric of Inquiry (POROI)* 12, no. 2 (2017): 1–13. https://doi.org/10.13008/2151-2957.1258.

———. "Rhetorics of Uncertainty: Networked Deliberations in Climate Risk." PhD diss., University of Arizona, 2015. ProQuest.

Walker, Kenneth, and Lauren Cagle. "Resilience Rhetorics in Science, Technology, and Medicine." *Project on Rhetoric of Inquiry (POROI)* 15, no. 1 (2020). https://ir.uiowa.edu/poroi.

Walker Kenneth, and Lynda Walsh. "'No One Knows What the Ultimate Consequences May Be': How Rachel Carson Transformed Scientific Uncertainty into a Site of Public Participation in *Silent Spring*." *Journal of Business and Technical Communication* 26, no. 1 (2012): 3–34.

Wallace-Wells, David. *The Uninhabitable Earth: Life after Warming.* New York: Penguin Random House, 2019.

Walsh, Lynda. "The Common Topoi of STEM Discourse: An Apologia and Methodological Proposal, with Pilot Survey." *Written Communication* 27, no. 1 (2010): 120–56.

———. "Resistance and Common Ground as Functions of Mis/aligned Attitudes: A Filter-Theory Analysis of Ranchers' Writings about the Mexican Wolf Blue Range Reintroduction Project." *Written Communication* 30, no. 4 (2013): 458–87.

———. *Scientists as Prophets: A Rhetorical Genealogy.* New York: Oxford University Press, 2013.

Walsh, Lynda, and Casey Boyle. *Topologies as Techniques for a Post-Critical Rhetoric.* New York: Palgrave-Macmillan, 2017.

Walsh, Lynda, et al. "Forum: Bruno Latour and Rhetoric." *Rhetoric Society Quarterly* 47, no. 5 (2017): 403–62.

Wanzer-Serrano, Darrel. *The New York Young Lords and the Struggle for Liberation.* Philadelphia: Temple University Press, 2015.

Watson, Matthew. "Derrida, Stengers, Latour, and Subalternist Cosmopolitics." *Theory, Culture, and Society* 31, no. 1 (2014): 75–98.

Wright, Lawrence. "America's Future Is Texas." *New Yorker*, July 10 and 17, 2017. www
.newyorker.com.

Ybarra-Frausto, Tomás. "Rasquachismo: a Chicano Sensibility." *Chicano Aesthetics:
Rasquachismo*. MARS, Movimiento Artiscico del Rio Salado, Phoenix, AZ, 1989.

Zimring, Carl. *Clean and White: A History of Environmental Racism in the United States*.
New York: New York University Press, 2016.

Žižek, Slavoj. "Slavoj Žižek on the Limits of Local Politics." November 19, 2016. https://
youtu.be/BEE51ay52Gc.

Index